SOCIAL STRUCTURE AND CHANGE

Volume 3

SOCIAL STRUCTURE AND CHANGE

Vol. 1. Theory and Method: An Evaluation of the Work of M. N. Srinivas (published 1996)
Vol. 2. **Women in** Indian Society (published 1996)

Forthcoming Volumes

Vol. 4. Development and Ethnicity
Vol. 5. Kinship and Religion

SOCIAL STRUCTURE AND CHANGE

Volume 3:

Complex Organizations and Urban Communities

EDITORS

A. M. SHAH
B. S. BAVISKAR
E. A. RAMASWAMY

SAGE PUBLICATIONS
New Delhi/Thousand Oaks/London

First published in 1996 by

Sage Publications India Pvt Ltd
M–32, Greater Kailash Market I
New Delhi 110 048

Sage Publications Inc
2455 Teller Road
Thousand Oaks, California 91320

Sage Publications Ltd
6 Bonhill Street
London EC2A 4PU

Published by Tejeshwar Singh for Sage Publications India Pvt Ltd, typeset by Line Arts Phototypesetters, Pondicherry and printed at Chaman Enterprises, Delhi.

Library of Congress Cataloging-in-Publication Data

Complex organizations and urban communities/editors A.M. Shah, B.S. Baviskar, E.A. Ramaswamy.

p. cm.—(Social structure and change: v. 3)
"In honour of M. N. Srinivas"—p. following t.p.
Includes bibliographical references and index.
1. Social classes—India. 2. Complex organizations—India. 3. Sociology, Urban—India. 4. India—Social conditions—1947– I. Shah, A. M. 1931– . II. Baviskar, B. S. (Baburao Shravan), 1931– . III. Ramaswamy, E. A. IV. Srinivas, Mysore Narasimhachar. V. Series.

HM15.S517 vol. 3 301—dc20 96–560
[HN690.Z9S6] [305.5'0954]

ISBN: 0–8039–9306–4 (Hb US) 81–7036–543–0 (Hb India)
0–8039–9307–2 (Pb US) 81–7036–544–9 (Pb India)

Sage Production Editors: Deepika Ganju and Evelyn George

In Honour of M. N. Srinivas

Contents

List of Tables

Preface

In 1982 several old students and friends of Professor M. N. Srinivas suggested that we organize a festschrift for him. His circle of students, colleagues and friends is so large and diverse in academic interests that we thought they could not be requested to write papers on any one particular theme. We therefore decided that we would request them to write on topics of their own choice, and then group the papers according to the major themes. The response was overwhelming. As many as 46 scholars promised to contribute papers. Some of them took more time than others in sending their papers. The process of commenting on the papers, returning them to the authors for revision, and editing the revised papers was far more time-consuming than we had anticipated. While some papers required little editing, others had to be edited in two or three stages. Only two authors did not respond to our comments, and therefore their papers could not be included. Since the time lag between the finalization of the papers and approaching the publisher was long, the authors were given an opportunity to update their papers if they so desired. A few of them took this opportunity to send entirely new papers; one paper was withdrawn at the last stage.

Considering the large number of papers, it was obvious that they could not be included in one volume. It was therefore decided to publish them in several volumes; however, the grouping of papers proved to be a difficult task. We tried several combinations and finally decided to publish the papers in five volumes. We then requested one contributor from each volume to write an introduction to it.

The papers in Volume 1 focussed on theory and method in the study of social structure and change, mainly through a critical examination of Professor Srinivas' work on India. Volume 2 focussed on various aspects of life of women in Indian society. The theme of Volume 3 is complex organizations and urban communities. Volume 4 will discuss various aspects of development and ethnicity; and the focus in Volume 5 will be on kinship and religion. The contributions in the five volumes cover a wide variety of structures and institutions and theoretical and methodological issues in the context of India as well as other societies. The papers presented here focus on a broad spectrum of themes in sociology and social anthropology which Professor Srinivas has either himself worked on or encouraged others to work on. We are honoured to dedicate these volumes to him.

We would like to thank Professor N. R. Sheth for writing the Introduction to this volume and for providing advice on editorial matters, and Shalini Suryanarayana for preparing the Index.

The secretarial expenses involved in our editorial work have been met from a small fund created by a few friends, and from grants made from time to time by the Centre of Advanced Study, Department of Sociology, University of Delhi. We are grateful to these friends and the Department.

A. M. Shah
B. S. Baviskar
E. A. Ramaswamy

Introduction

N. R. SHETH

I

This volume, like its companions, is dedicated to the scholarly and professional excellence of M. N. Srinivas and the rich treasure of academic wealth he has produced during his career of over four decades as teacher, researcher, writer and administrator. Most of the contributors in this volume have been his students or colleagues at various stages of their career development. I, for one, had the unique privilege of being one of Srinivas' first undergraduate students. My 'remembered village' of learning sociology includes personalized discourses on understanding one's own society by studying others, recognition of the splendour and value of sensitively scientific observation of social reality, and appreciation of the great tradition of enriching scholarship by humanism and humour. I also learnt the importance of writing simple yet elegant English for precise and effective communication. Many of my friends and colleagues know how the Professor's scribblings on my written drafts helped to enrich my academic, intellectual and communicative assets. I am sure other students of Srinivas were also recipients of similar bounty.

As a scholar brought up in the functionalist perspective of British social anthropology, Srinivas spent most of his academic career trying to unravel the intricate maze of the foundational institutions of India, such as caste, religion, kinship and the rural community. His preoccupation with local institutions has, however, been informed by an abiding interest in the *gestalt* of contemporary Indian social reality, in which the classical and contemporary dissolve into each other. Accordingly, even as he pursued and promoted social anthropological research among rural and tribal communities, he soon recognized the value of studying the urban-industrial phenomenon in India by the fieldwork method. His close contact with the emerging trends in social anthropology in the West made him aware that the use of the tools of social anthropology for understanding complex organizations had begun to yield good results in spite of its obvious limitations. For instance, the well-known Hawthorne studies of human behaviour in an industrial organization under the leadership of Elton Mayo had led to an almost revolutionary insight into the reality of management-worker relationships and the problems of productivity and performance in economic enterprise. In a similar perspective, Tom Lupton studied shop-floor relations in a British factory by using the fieldwork method of social anthropology. With his knowledge of such developments, Srinivas was convinced that the distinctive methodological assets of social anthropology could be used to gain insights into the social dimensions of enterprise and performance in the Indian context. He, therefore, inspired some of his students to undertake studies of the various segments of urban and industrial reality. His students, and in turn their students, have contributed to teaching and research in urban, industrial and organizational sociology, involving a spectrum of subjects of contemporary value (such as industrial leadership, management, trade unions, urban communities, and cooperatives). His knowledge and wisdom were utilized by government and non-government agencies from time to time to deal with complex socio-economic issues. In 1967 the National Commission on Labour requested him to chair a study-group on the sociological aspects of labour-management relations as they were convinced that his scholarly insights into Indian social reality would provide a fresh perspective to the examination of industrial relations. The contributions presented in this volume are a tribute to such a holistic intellectual involvement in Indian society and an incisive yet firmly grounded view of how social reality should be approached as a subject of research.

II

The theme of this volume is complex organizations and urban society. The ten contributions presented here cover a wide and heterogeneous range of sociological analysis. Six papers (by Ramaswamy, Panini, Newell, Bailey, Barnabas and Sheth) deal with the behaviour patterns and processes of people who play important roles in complex organizations. The others (by Gupta, D'Souza and Gill, Damle and Trivedi) are concerned with the social aspects of urban communities and the process of urbanization.

Ramaswamy, Panini and Newell have analyzed the behaviour and attitudes of people in Indian industry to show the interplay of the rational-bureaucratic and socio-cultural factors in formal organizations. Ramaswamy has used his rich and varied exposure to the employment relations scene in India to bring out the pervasive influence of the idioms of traditional Indian culture on the spectrum of labour-management interactions. In an unemployment-ridden social system, only a small proportion of job-seekers succeed in getting regular employment. These privileged people then aspire for job security at any cost to the employer, and for other benefits like easy promotions, status-lending designations, employment guarantee for their progeny, and the use of trade union power for their sectional advantage. Consequently, work organizations are overmanned and manual work is frowned upon. Organizations are loaded with too many officers and supervisors gloating about their status and privileges but with little power or control over work. Managers, ironically, want legal recognition as workers for the sake of job security. Trade unions are fragmented in pursuit of sectional interests. The rationality of a modern production enterprise is bounded by the fractional rationality of employees seeking immediate gains to enhance their status within and outside the enterprise. The logic of wealth, power and control over the work situation is converted into the logic of status which, Ramaswamy argues, is based on the predominating value of status in traditional Indian society.

Panini is concerned with industrial entrepreneurs who own and/or manage complex organizations in an environment of changing technology, increasing competitive markets and economic uncertainty. At the lower levels, worker-entrepreneurs with modest education and low economic aspirations follow the rationality of a business enterprise in a pragmatic sense, guided by the goal of 'decent living with hard work'. At the same time, they retain strong links with the old institutional values of family, caste and religious faith. As you move up the scale of entrepreneurship, the old institutional loyalties and values are retained, but they are subordinated to the techno-economic logic of modern enterprise governed by reflective rationality (as against the *ad-hoc* rationality of small entrepreneurs). At the highest level, people with high economic and social stakes in organizations use old loyalties to provide social, economic and political patronage, which is useful to them in achieving the objectives of their enterprise. In the organizational context, the bonds of kinship, caste, religion, and so on, mesh into the rationality of the enterprise. The continuance of primordial loyalties in the world outside does not act as a hindrance to people's performance in their organizational roles.

The take-off point of Newell's paper is the mounting evidence that traditional institutions like caste do not have much influence on the social behaviour and attitudes of industrial workers at the workplace. Newell's study of workers in the Kanpur textile industry leads him to conclude that while caste does not survive in its old form in industry, one of its major attributes, namely, hierarchy of status, pervades social interactions among workers. The hierarchy of status corresponds to the hierarchy of job categories, with the concomitant hierarchy of skill, earnings and cultural nuances of dress, language, and so on. Newell observes casteist inflexibility and divisiveness among workers at various levels. Even union leadership lends status, and hence is coveted by those who cannot hope to raise their status in the job hierarchy. Newell seems to be convinced that caste-type social divisions and hierarchy characterize the pattern of social interactions among people in Indian industry.

Barnabas has examined the attitude of administrators in Central government services to change. Contrary to popular belief and classical sociological wisdom, he finds that Indian bureaucrats are positively oriented to change, and that the level of education acts as a positive force in a bureaucrat's attitude to change. This implies that complex organizations in the government sector are governed by people who can act as change agents. An extrapolation relevant to our theme is that administrators with relatively high education may not be restricted in their actions and decisions by the controls of traditional institutions.

Bailey provides valuable insights on how the multiplicity of goals in a complex organization reflects the behaviour and attitudes of the various sections of its members, and how these members mould the diversity which affects the organization's performance. Teachers in a university bring their ideas and preferences about the various goals from their *alma mater*. The consequent disparity of perceived goals among sections of teachers contributes to dissensus about the major tasks of the university. They exert pressure on the organization to direct the thrust of work towards their preferred goals. This process sharpens the dissensus and affects the performance of the university. Non-achievement of goals in turn increases dissensus among members, creating a vicious circle. The fewer the goals involved in diversity, the sharper the dissensus among the members. Bailey suggests that the resolution of conflicting goals should be based on a thorough understanding of the consistency between what people say and what they think. The observed dissensus is a function of the political process of stakeholding in an organization, and the strategy adopted by people to push their ideas and values.

Sheth highlights the importance of the conflict perspective as an analytical aid in the study of formal organizations. He explains the relevance of this perspective in sociological research in the context of the growing recognition of the need for closer ties between sociology and social governance. Conflict in an organization is a product of its inherent political structure and process. It is necessary to study how power is acquired and used by various categories of members in an enterprise according to their stake in it. This would help to understand how conflicts are generated and handled at various levels. Such an understanding would make sociological analysis more relevant and meaningful in the purposive comprehension of the constellation of social forces comprising social organizations.

Gupta's study of an urban community of middlemen traders illustrates the emergence of complex organizations in the area of marketing of agricultural produce. The entry of farmers and traders into the modern market economy converts their relations from the old patron-client (*jajmani*) bonds to more formal and impersonal ties of economic calculus, enabling the traders to buy and sell on an opportunistic basis. This leads to the decline of trust and social interaction between the trader and the farmer, and gives considerable power to the trader to exploit the farmer's economic vulnerability. While the government seeks to control the farmer's exploitation by regulatory and supervisory authority, the trader uses his vicarious economic power to neutralize the government's authority. This potentially engenders a process of conflict between the traders and farmers.

D'Souza and Gill have analysed the process of emergence of a trading community in an urban centre. They have highlighted the role of caste as a restrictive force in the choice of trade by people who migrate from rural areas essentially to avail of new economic opportunities. Unlike large industrial centres, the freedom to choose entrepreneurial openings in small trading centres is limited by the norms and sanctions of traditional institutions such as caste.

Damle demonstrates how the occupants of various types of neighbourhood structures in an urban centre adapt to economic, cultural and social modernization. The acceptance of Western ideas, values and social interaction does not destroy people's allegiance to the old values and institutions. The forces of modernity contained in the inevitable process of change are absorbed by the traditional institutions and values. The fusion of the two results in the development of a new pattern of social relationships.

Trivedi highlights the limitation of distinguishing rural and urban societies as two contrasting entities. While secondary relations are predominant in urban areas, urban people do not give up their allegiance to primary relations easily. In a developing society, the urban community needs to develop a rural orientation in view of the continuing presence of organic links between the urban and the rural. In response to this need, Trivedi argues, an intermediate social entity called a semi-urban pocket (SUP) has emerged. He presents the main attributes of SUPs in urban and rural environments. While SUPs in developed societies reflect the process of traditionalization of modernity, they represent the modernization of tradition in developing societies. Trivedi's analysis involves a wide and diverse range of issues on urbanization and urban sociology.

III

A major concern of the scholars contributing to this volume centres around the role played by traditional Indian institutions, especially caste, in the social behaviour and attitudes of people participating in today's complex organizations and communities. It is explicitly or implicitly assumed that people's involvement in or allegiance to primordial Indian institutions is a discrete sociological phenomenon in relation to their participation in complex organizations or urban communities, which are characterised by impersonal secondary relations, formal rules and sanctions and the techno-economic logic of performance and productivity. Panini's research suggests that entrepreneurs can maintain their primordial loyalties as well as comply with the logic of modern enterprise without any significant incompatibility between the two. Likewise, Damle shows that the impact of modern educational and occupational culture on urban conglomerations does not wipe out their inhabitants' primordial loyalties. The old and new loyalties are intertwined to create a new pattern of social relations.

On the other hand, D'Souza and Gill have shown that the occupational and social disabilities inherent in the traditional caste system follow the rural migrants to a new urban trading centre and create unequal economic opportunities. A similar handicap is found by Newell in the industrial sector. He shows how caste acts as a restrictive force in the distribution of job opportunities in the first place. While caste as such is not significant in the network of workplace relations, the crucial element of status in the caste system influences the behaviour of workers in the job hierarchy. Workers in each category maintain social and cultural exclusiveness. Mobility is either absent or restricted at the lower levels of skill and status. Ramaswamy, on a wider plane of analysis, also underscores the impact of the attribute of status on the behaviour and performance of people in industry. In his view, preoccupation with status pushes other attributes of social hierarchy (wealth and power) into the background and induces people to exploit existing opportunities and privileges to grab more and more status symbols at the cost of the formal rationality of modern enterprise.

As the various papers in this volume are obviously not governed by a common conceptual or analytical framework, the similarities and diversities in their findings and conclusions are mainly coincidental. However, they bring home the crucial lesson that it is fruitless to look for a common pattern across India in the search for an indigenous socio-cultural profile of contemporary complex organizations. The social and cultural processes within an organization need to be observed with reference to the multitude of internal and environmental forces influencing its structure and performance—forces such as its bureaucratic system, technology, economic environment, interface with the government and wider society, and the mix of primordial bonds and loyalties observed by its members at various levels.

A bias in favour of such a comprehensive approach induced me to plead in my paper for a purposive sociology of organizations which should move beyond the conventional functionalist-integrationist perspective. If we look at the sociological phenomenon of an organization on the basic premise that dissensus and conflict are as endemic as consensus and cooperation in any segment of society, we may be able to understand the phenomenon in sufficient depth. Apart from the various sources and uses of power and its role in the formation of conflict groups, as I have argued, we should also examine the ways in which the abuse of power for selfish or sectional ends generates countervailing power among conflict groups and the interplay of forces wielding official and countervailing power. This would mean that we should resist the intellectual temptation to consider identification or quantification of the role of primordial institutions as the primary task of the sociology of formal organizations. In practice, those scholars who argue that primordial loyalties do not impair the logic of modern enterprise, as well as those who regard such loyalties as detrimental to the logic of enterprise, subscribe to the assumption of traditional bonds (like caste) as discrete entities which can be superimposed on the organizational structure and process. We need to abandon this conventionally charming but not-so-productive path of sociological inquiry. Instead, we should assume that the sociologist's primary task in studying complex organizational or urban phenomena is to comprehend the totality of social relations, or any segment of that totality, in a purposive manner. Within such a framework, our scholarly effort may yield more relevant and useful results.

The acceptance of this approach in the sphere of urban sociology would mean moving away from the attractive tendency to equate 'urban' with 'complex', 'modern', 'industrial' and other such conceptual abstractions. While there is obviously much to be said in favour of such equations in respect of contemporary urban agglomerations everywhere, it is of extremely limited sociological value to deal with the urban-rural distinction as a manifestation of a dichotomy or continuum. A specific urban reality should be examined as a discrete entity which may contain segments of the primordial and secondary relations in a blended form. In this context, it will be interesting for some scholars to study a spectrum of urban communities within the framework of Trivedi's belaboured postulate of the semi-urban pocket as an interface between the urban and the rural. I suspect that any segment of urban or rural reality may show degrees of what Trivedi would call semi-urban. We should therefore develop greater analytical interest in the blended reality of social relations in a given situation than in conventional distinctive categories such as urban and rural.

Among the articles on complex organizations, Bailey's contribution represents a remarkable effort to understand the totality of social relations with a well-defined purpose. Today's complex organizations are almost by definition designed and managed with a variety of goals. As the goals are often incompatible, a consensus on goals is vital for smooth and effective performance. At the same time, dissensus on goals is inevitable as all members of the organization bring into it their respective predispositions based on their extra-organizational affiliations and loyalties. Bailey unfolds the process of conversion of dissensus into the crucial consensus and, in doing so, provides some valuable insights into the magnitude and quality of dissensus about a university's goals among members of its staff. Dissensus needs to be differentiated in terms of its location in people's thoughts, words and actions. The sociological value of dissensus increases with the degree of failure in the achievement of goals. People disagree about whether dissensus represents institutional sickness. Dissensus can be reduced in several ways. But if dissensus is reduced to fewer goals, the intensity of conflict among members may increase. In substantive terms, the intellectually and pragmatically attractive search for central values acceptable to all members may not help to eliminate or minimize dissensus. Consistency and consensus on organizational goals should be regarded as a political process depending on people's stakes in the organization and the strategy adopted by them to live through it.

I have dwelt here on Bailey's contribution at some length at the cost of some repetition because his approach and analysis, in my view, hold sociological value across cultures. Its value may be limited across organizational types as the members of a university enjoy a collegiate culture which may not exist in industry and other more bureaucratic organizations. However, with the winds of liberalization and individualism sweeping social systems across the globe, it seems complex organizations are increasingly going to face problems of inconsistencies, dissensus and conflict among their stakeholders. Bailey's argument is, therefore, likely to be relevant in a growing variety of organizational types.

Bailey's analysis of dissensus on organizational goals includes a question about how dissensus enters an institution. He identifies the '*alma mater*' factor as a principal cause of dissensus. University teachers bring with them philosophies, ideas and assumptions from their respective schools about what a university should be and do. These philosophies, ideas and assumptions shape their preferences and prejudices relating to university goals and consequently the nature and extent of dissensus on goals. This part of the analysis bears some similarity to the studies which look for the role of primordial institutions in shaping the behaviour of people in Indian industry. However, in the scheme of things represented by Bailey's study, you do not begin by imputing a major role to primordial institutions (in his case, the *alma mater*) and then discover such a role; instead, you begin by looking for behavioural facts in their organizational totality and bring in primordial loyalties where they assist you in explaining some segment or segments of organizational behaviour.

Ramaswamy's paper constitutes a welcome effort in such an approach to organizational sociology. His main concern in the essay is to explore the effect of Indian society on Indian industry. However, his exploration is guided by the vantage point of the multisplendoured social reality in industry. In a labour market afflicted with unemployment, people who secure jobs in industry treat it as a haven of economic and social security. Hence recruitment to permanent jobs is often a function of political or anti-social pressures on the employers rather than of genuine need of an organization. Having attained the status of an industrial worker with all its legal and social privileges, workers constantly strive to enhance their status. Other symbols of social differentiation, such as wealth and power, are used or manipulated to support the aspiration for status. Ramaswamy regards the conversion of wealth and power into status as the most pervasive social process in industry. The constant search for status is viewed by him as a fundamental social value, rooted in the traditional caste system and analogous to the process of sanskritization. Ramaswamy is careful not to pose his reflections on the value of status as a universally applicable model in the Indian context. But his findings should serve as a welcome challenge in the progress of organizational sociology in India. We need to discover variations in the process of conversion of wealth and power into status across organizational types and hierarchies. In traditional Indian society, it was perhaps not uncommon for social groups to look for political advantage when avenues for raising status were closed by superimposing disabilities such as untouchability. In any case, the hypothesis regarding the centrality of status in Indian society and industry needs closer examination in sociological research. The search for status and the need to convert wealth and power into status may be of crucial importance at the middle and lower levels of organizational hierarchy. However, at the higher levels, bureaucrats, professionals and entrepreneurs may regard status and wealth as instruments of control over resources or simply for the achievement of the goals of performance and productivity.

The value of status as a principal object of social differentiation and aspiration in present Indian society may be a function of growing social, economic and political inequalities in the wake of ever-rising population pressures, widening disparities of income and opportunity, and the mounting invasion of material goods and services generated by the global consumerist culture. As citizens of a poor democracy surrounded by the products of modern technology, we are perhaps being brainwashed into judging our social standing and being judged by others in terms of our hold over material symbols of comfort and gratification. As the availability and value of material artefacts of the new civilization seem to be spreading among contemporary societies, the predominance of status as an axis of social differentiation and ambition may grow across socio-cultural boundaries. Status as a social value, as illustrated by Ramaswamy in the modern Indian context, may then be on the verge of attaining a global stature.

IV

Japan's experience of industrialization has made veteran social scientists demolish the old Western myth that there is a single cultural path to the achievement of targeted performance in modern complex organizations. Indeed, the unique attributes of Japanese culture, such as permanent loyalty to an organization and a strong affective element in employer-employee relations, which could be easily reckoned as obstructive to productivity by conventional social science wisdom, have been used in Japan as positive inputs in industrial performance. Can we follow them and turn seemingly counterproductive cultural qualities into productive assets? If pursuit of status occupies a predominant role in Indian tradition, as Ramaswamy argues, we can share his optimistic proclivity. We need to find ways of converting the social value of status from its present obstructive role into a valuable asset for organizational performance. This is a major challenge for organizational analysis and practice.

However, the social significance of status is only one of the unique attributes of Indian society proposed as a potential asset to modern complex organizations. There is by now a significant amount of discourse in social science and management literature on the possible contribution of unique Indian values to the performance of modern enterprise. Some scholars stress the importance of Indian familial values of protective paternalism and respect for age and experience. It is contended that such values may contribute to stress and anxiety reduction in superior-subordinate relations and promote mutual understanding and loyalty, with positive effects on performance at the workplace. Similarly, there is growing opinion among social scientists that ancient Indian spiritual and philosophical values can become major assets in raising the levels of performance and quality of work-life in modern enterprise. This opinion is vigorously promoted by Indian and Western scholars who are concerned about the rapidly degrading social and natural environment in the wake of the worldwide ascent of individualism, liberalism, consumerism and normlessness.[1]

While this search for culturally specific or unique models of complex organizations will continue and perhaps grow, we cannot wish away or underestimate the overarching influence of cultural globalization. The emerging concept of the world being a village is all but a social reality. The recent rapid disintegration of the socialist mode of economic and political governance has brought the Western capitalist system to the forefront as a coveted method of managing social enterprise at all levels. This implies that the enterprise will be predominantly guided by liberal-democratic and individualistic values. The process of globalization of markets and the economy will be reinforced and hastened with the use of today's quick and effective methods of communication and transport. Dissemination of information around the world by the modern media of communication tends to create a homogeneity in ideas, ideologies and values across the world. Is there a poetic justice in social science which will lead organizations around the world to move in tandem towards a new, emerging global culture? Such a prospect may bring the old convergence

[1] A comprehensive and sustained analysis of the relevance of Indian scriptural and institutional values for modern organizations and the application of these values in concrete organizational practice is made by Chakraborty (1991).

theory (which predicted the convergence of all industrializing societies towards a uniform socio-cultural pattern) down from the shelf, although its protagonists may not ever have dreamt of today's 'high-tech' global culture.

It is, of course, too early to foresee when or in what way the emerging global culture will acquire the critical mass to act as a leveller of regional and local cultures. But one can already see some straws in the wind blowing, interestingly, from Japan. It is now known that the unique cultural traits of employer-employee relations adored by social scientists and management theorists since the sixties were not practised uniformly in Japanese industry. They were more relevant to assembly-line technologies than to others. At the same time, as Japan progressively got involved in international enterprise, trade and economy, its people began to get acquainted with the globalizing values of individualism and the preference of merit over seniority. In recent years, the frequent economic crises faced by the country as a member of the club of super powers have convinced observers and social scientists that Japanese industry should adopt Western values of work and human relations to be able to survive and move honourably in the global economy.[2] India's role in the global economy is vastly different from that of Japan. India is socially too heterogeneous and economically too poor to provide a clear and meaningful image of the impact of the emerging global culture on its organizational systems and achievements. Moreover, the techno-economic compulsions imposed by the forces upholding the global economy on the one hand, and the social and economic constraints dictated by a labour market infested with unemployment and poverty on the other hand, generate conflicting pressures on Indian enterprise. The intricate maze of economic forces may cause a widening distance between large and small, rich and poor, stable and unstable, and technology-intensive and labour-intensive organizations. Those at the upper end of the techno-economic divide may be drawn more towards the new global culture than those at the bottom. But all organizations may contain or adopt some elements of the local culture. In fact, it is likely that all

[2] By an intriguing cultural and intellectual irony, while social scientists and management scholars outside Japan continue to promote the Japanese style of management as an alternative to the Western models, Japanese scholars have increasingly been stressing the limitations of this model and the inevitability of Japan adopting the Western model. See, for instance, Kobayashi (1990). Another brief but clear message on this subject is contained in Miyai (1992).

organizations would go through a twin process of globalization with localization. The sociology of complex organizations may have to attend increasingly to this dual trend in the years ahead.

The growing tempo of economic liberalization and the limits of the government's ability to promote and manage enterprise have begun to create a proliferation in the number of organizations, especially in services such as health, education and rural development. An increasing number of voluntary non-government agencies are adopting the role of fulfilling socio-economic objectives in the service sector. Such organizations need to be designed discretely for their respective specific purposes. Most of these organizations are designed with a strong commitment to the values of individual freedom and as open, adaptive and dynamic structures. They also adopt a participative management style and a commitment to the continuing development of human resources as an integral part of organizational success and development. In such efforts at designing and managing complex organizations, the strength and commitment of leaders occupy a central place. In today's turbulent environment, we need to identify leadership patterns and structures which would make organizations not only stable and successful but also creative and innovative.

A growing tendency is visible among leading managers and entrepreneurs to look for a specific set of indigenous values to be introduced into organizational design or practice. Chakraborty's scholastic and practical efforts, which I have mentioned in the foregoing, represent the introduction of Indian values by intervention from outsiders. In other cases, the organization's leaders themselves develop commitment to some values and adopt these values as a guiding force in managing its activities. In one unit, for instance, the chief executive introduced the model of the extended family in managing the enterprise and inspired his colleagues at all levels to treat each other as kin. At the same time, the employees were encouraged and often cajoled into free discussion on all important issues and decision-making by consensus. Blackboards and other writing facilities were installed in various parts of the unit's premises to aid discussion and expression of dissent. Thus the values characterizing both family and school were adopted. It was clearly understood that the incorporation of such values would lead to the more effective attainment of the main organizational goals. In another organization, the members of work groups were required to rededicate themselves each morning to the principal service-oriented goals of the enterprise and also to the observance of purity of means in the true Gandhian spirit. Here the organization was treated like a place of worship.[3] Such selective use of indigenous values in the management of complex organizations and their impact on the overall network of relations within the enterprise are subjects of considerable sociological value. In the course of its life, each organization probably goes through the process of constructing its own culture with ingredients partly borrowed from the outside world and partly manufactured within. This process supplies a well-endowed field for the sociologist to discover.

Since the days of Elton Mayo, who inspired and guided the pioneering Hawthorne research in human aspects of industry, the predominant thrust of organizational sociology is to place the human being in the centre-stage of all systems and activities. Social science, in this sense, has worked for the development and growth of a culture of humanism in modern organizations. Organizations and their performance are regarded as a means to the end of human satisfaction and happiness, so we say or mean. All modern humane or progressive methods of dealing with people, such as participative management, the quality of work-life, humanization of the work environment and human resource development are designed to carry us forward towards the main human objectives. How far have we travelled towards these objectives? This is a confounding question. It is difficult to say how 'humane' organizations have become since the birth of the human relations approach. One could, without fear of contradiction, make a significant negative statement that, in the absence of that approach, organizations would have become more coercive than they are at present. However, organizations are still disturbingly exploitative of their human ingredient. This ingredient, of course, is not a uniform entity. Most of the coercion of human beings by organizations is indeed coercion of one set of human beings by another, Marx or no Marx. Some humans are coerced by organizations while others indulge in coercing organizations for selfish ends. The social reality of organizations often makes us aware that, paradoxically, the more an organization tries to change towards a culture of participative management and humanization, the more individuals and sections of its membership

[3] Information on the two organizations presented here is based on unpublished documentation of cases in management innovation for a research project sponsored by the International Management Development Network. I am grateful to my colleagues Deepti Bhatnagar and Mukund Dixit who are involved in this project and have given me some useful insights.

experience coercion. For instance, an increasing number of Indian organizations in recent years have adopted the new humanistic philosophy and approach of human resource development. On the other hand, the mounting economic crisis in industry has pushed organizations to the wall, forced managements to reduce their workforce and introduce stricter regimes of productivity on the shop floor. An important part of the agenda of the sociology of complex organizations consists of a compassionate understanding of such paradoxes and incongruities as I have just illustrated. This kind of agenda may remain unfinished for long, regardless of whether industrial societies converge or diverge. To accomplish this task, sociologists need a lot of the qualities which Bailey admires in Srinivas—an ever-present sense of empirical reality and emphasis on understanding diversity, not intellectual excitement.

REFERENCES

Chakraborty, S. K. 1991. *Management by Values: Towards Cultural Congruence*. Delhi: Oxford University Press.

Kobayashi, N. 1990. 'The Response of Management Education to the Challenges Facing Japanese Business in the Twenty-first Century'. In Max Von Zur Muehlen (Ed.), *The Search for Global Management: Mapping the Future of Management Education and Development*. Geneva: CFDMAS and Interman, pp. 21–30.

Miyai, J. 1992. *Japanese Approach to Human Resource Development*. Tokyo: Japan Productivity Centre.

1

Wealth and Power Convert Into Status: The Impact of Society on Industry

E. A. RAMASWAMY

CONVERGENCE THEORIES

It is now three decades since a group of influential labour economists led by Clark Kerr came up with their far-reaching hypothesis on what industry would do to society (see Kerr et al. 1973). As industrialization becomes the engine of economic development, they claimed, societies would find it necessary to foster beliefs, values and attitudes which are consistent with, and conducive to, the industrial way of life. Industrial societies across the globe were expected, under the compulsion of this economic logic, to move away from cultural differences and converge towards a set of common values. The years which have rolled by have done little to sustain this bold prediction. Countries with diverse cultures which have entered the race to industrialize have brought to the process vastly different beliefs and value systems. Witness, for example, the methods of work organization and human resource management which are thought to underlie Japan's industrial might. While Western society thought of the unencumbered mobility of labour in search of better compensation and affectively neutral relationships between employer and employee as the quintessential industrial values, Japan built its industrial empire on employee commitment to the enterprise promoted by life-time employment policies and affective relationships at the place of work. Japan's thrust towards decentralized decision-making and flexible work organization come once again as a contrast to Western society's faith in Weberian bureaucracy and Taylorian division of labour.

The superiority of Japanese work organization led Ronald Dore to propose a reverse convergence—a situation where Western industrial society is forced to adopt the 'organization-oriented system' of Japan instead of the rest of the world having to shift towards the 'market-oriented system' of the West. When first suggested in the mid-seventies in his *British Factory—Japanese Factory* (1973), Dore's theory of reverse convergence, as he himself has observed more recently (1989)

> seemed a bit bizarre ... Lord McCarthy...clearly found it offensive; 'naive and fallacious' were his exact words. Robert Cole thought it 'muddied' the otherwise potable waters of the book; 'that Japan is further along in the evolutionary scheme of development with its brand of management-labour cooperation is sheer fantasy—and not necessarily a pleasant one'.

Two decades after Dore's book was published, it is easy to see which way the pendulum has swung—Anglo-Saxon 'market-orientation' has taken a beating at the hands of Japanese 'organization-orientation'.

Even as the rest of the world is trying to catch up with Japan by imitating its human resource strategies, there is growing evidence that Japan itself may be shifting away from such practices as seniority-linked promotions and salaries and life-time commitment to the firm. Moreover, other industrial success stories are beginning to emerge, especially in the Pacific rim. Taiwan, South Korea and Singapore are already being reckoned as important industrial powers, with Indonesia and Malaysia not far behind. We know very little as yet about the values underpinning their industrialization, which might well be different from those of both Japan and the West. China might emerge as the industrial leviathan of the twenty-first century, further upsetting neat little theories about social values appropriate to successful industrialization.

India hardly counts in the league of industrial nations. Although large-scale factory production may have existed for over a century, our industrial base is too narrow, industrial employment far too limited and the contribution of industry to the overall economy too meagre for us to be counted as an industrial society. Our experimentation with industrialization can only be described as halting and fitful, especially because in the crucial years since independence industry had become the testing ground for doomed ideologies. In the midst of such dismal performance, there is growing evidence that industry does not operate in a social vacuum. The last two decades would in particular count as a period of far-reaching influence of society upon industry, although the process is largely undocumented and even unnoticed. This paper is about this impact, about the demands society places on industry, the pressures it brings to bear, and the challenges it poses to both labour and management. As must be the case with every society, there is much about this interface between industry and society which is uniquely Indian. A model of industrialization which would hopefully produce better results than in the past would have to reckon with these social forces. In putting this account together, I have drawn on my experience with labour and management in large-scale industry, especially during the last decade which has been a period of intensive involvement in many different capacities—as teacher, trainer, researcher, consultant and columnist. It is a generalized picture made up of bits and pieces from many different situations, and not a depiction of the scene in any particular enterprise or region.

THE SEARCH FOR EMPLOYMENT: WHO GETS IN AND WHY?

The social forces that bear on industry can be traced in many ways to the aspirations of people to whose fulfilment industrial employment is the perceived vehicle. Work is a means of earning a livelihood, but this requirement is met rather easily because industry pays well, often exceedingly well, in comparison with what those outside its portals get. Compensation is attractive although employment is scarce, and getting on to the rolls of permanent labour is becoming more and more difficult. Job security comes easily as well. With retrenchment of surplus labour and closure of losing concerns so hemmed in by restrictions as to be made virtually impossible, any permanent worker who keeps away from serious misconduct has a cast-iron guarantee of life-time employment.[1]

Since employment on such generous terms is the objective of vast numbers of job-seekers, industry has become what Holmstrom (1976) calls a citadel. Inside the citadel are a small handful—some call them an elite or, even more disapprovingly, a labour aristocracy—enjoying these enviable benefits, while vast hordes storm it from outside in the hope of breaking in. Who has claim to the few jobs that industry has to offer? The list is unbelievably long. 'Land losers' whose lands were appropriated to build the plant have a natural claim, followed closely by other indigenes, although none of them may have the skills industry wants. Massive industrial projects in coal, steel, fertilizer and power generation have, for instance, been located in backward areas in the interests of regional development or because of raw material availability. Not a few of them employ such sophisticated technology as to dispense with unskilled labour, but the population of a tribal tract has nothing else to offer. The wholesale import of skilled labour may be the most rational policy, but hardly feasible, especially if a public enterprise has come up on the site.[2] Even the private sector is realising that it must live in harmony with the immediate society, and that means taking on board some local people whether or not they have the appropriate skills. On one of my trips to a factory some 50 miles from Calcutta, the general manager with whom I was travelling had

[1] With the liberalization of the economy, employers are demanding an exit policy, which is widely understood as the loosening of controls over the retrenchment of unwanted employees and the closure of unprofitable businesses. When, and even whether, the government will amend the laws is yet unclear. For now, Indian workers enjoy unparalleled employment security. For a discussion of employment security in India, see Mathur (1989: 153–202).

[2] I have witnessed local politicians in a West Bengal public sector construction site insist that contractors provide employment to their party supporters if they wanted to retain their contracts.

a gunman seated in the front of the car. It was impossible to travel without armed escort, the manager complained, because people stopped him all along the route to ask for food. They thought he had the power to give them jobs, which was unfortunately not the case. Industry is veering to the view that it is better to take on unskilled local people and teach them skills than keep them out and risk the consequences.

The claim of the local population on industry does not stop at employment, however. Does a woman from an adjacent village in the throes of labour pain have any claim on the factory hospital? I have been asked this very question by personnel managers in training sessions. While factory hospitals are responsible for the health of employees and not of the community, how should the doctor react to an emergency when his is the only hospital within a 50 kilometre radius? And what should schools set up to educate employees' children do with parents from the neighbourhood who want their children educated? Hospitals, schools, water supply and housing built by industry for its employees often come as a plush contrast to the desolation of the surrounding countryside. While these facilities can be neither thrown open to the general public nor denied altogether, deciding where to draw the line is far from easy. Industry is discovering, however, that it must go beyond the conventional balm to the conscience. Drilling a couple of wells or despatching an ambulance to the neighbouring village and proclaiming in the last paragraph of the chairman's address to shareholders that the company's social responsibility has been discharged is not going to suffice much longer.

Witness, for example, the experience of a tobacco company with the People's War Group—a militant outfit pledged to the liquidation of class enemies—which has sway over the countryside in its backyard. Managers have been summoned by the group to desolate spots in the dead of the night and questioned about a range of issues from employment to wages, bonuses, and confirmation of temporary hands. The firm would have been forced to wind up operations except for the goodwill of farmers earned through years of extension work on promoting tobacco cultivation and providing agricultural inputs. The local community has become a powerful buffer insulating the firm from the depredations of the militant outfit.

Coming back to employment, the sons-of-the-soil are just one of the many claimants to jobs. Labourers engaged to construct the buildings and erect the machinery, most of whom are once again unskilled, will not pack their bags and leave once their work is over. Local heavyweights of all kinds (from members of parliament and the legislative assembly to opinion leaders, the police, civil administration, freedom-fighters and such other influentials) will push their candidates. Scheduled Castes, Scheduled Tribes, physically handicapped and the like will demand employment by way of affirmative action.

While greenfield operations battle with these competing claims, the problems of running plants are scarcely less complex. With jobs so highly prized, there is hardly a trade union which will not use its bargaining power to demand employment for its members' offspring. Even though such preferentially recruited manpower is often of poor quality, there are literally scores of collective agreements in public, private and multinational enterprises setting aside a portion of future vacancies for employees' heirs.[3] Every firm has, in addition, layers of less privileged workers (as for example contract labourers, casuals and *badlis*), all of whom want to get on to the permanent rolls. Experiments with alternatives to permanent employment have met with limited success. In what must be reckoned as an especially astute move, a petroleum refinery decided to replace its labour contractors with a cooperative consisting of their labourers. With the refinery assuring a steady flow of work and deputing some of its own executives to manage the cooperative, dramatic improvements were achieved in wages, benefits and job security. When rapacious contractors became a distant memory, however, the members of the cooperative formed a trade union whose chief demand was permanent employment in the refinery.[4]

[3] A novelty when Uma Ramaswamy (1983) reported its presence in the Coimbatore textile industry, this practice can now be found in every part of the country and every kind of enterprise. Coal India, with over 750,000 employees, is a prime example in the public sector. Examples in the private sector include such well known names as Hindustan Lever, Boehringer Knoll and Vazir Sultan Tobacco. Workers use this right to settle the least capable offspring, since the more able ones can fend for themselves. There are also cases where the right is hawked to the highest bidder because a worker may have no offspring or may not wish to nominate one. This explains the poor quality of labour recruited through this route.

[4] Self-employment is a somewhat more successful alternative, though not for all categories of workers. Some firms have pursued this route to set up enterprising workers and middle managers as suppliers. Skilled workers of engineering firms have been given the option to buy off the very lathes or machine tools on which they work and turn sub-contractor. Drivers have similarly turned into transport operators plying the same cars for the company.

Claimants to jobs in organized industry are so many and so varied that personnel managers charged with manpower recruitment spend their energies balancing competing claims rather than matching labour supply with the firm's own requirement. While industry cannot meet all of the claims, it cannot ignore them all either. Formal rationality might demand a lean and trim labour force finely tuned to requirement, but substantive rationality dictates a compromise which will help industry mesh with its environment. What industry does—and some firms do it a lot better than others—is to selectively yield to pressures without altogether losing sight of actual requirement. The obvious consequence is overmanning, which is endemic to the organized sector and especially chronic in public enterprises. It can hardly be otherwise, however, given the compulsions of the environment in which industry operates.

THE QUEST FOR MOBILITY: LEVELS, DESIGNATIONS AND PERQUISITES

While those outside industry want to get in, an important objective of the ones already inside is social mobility. The quest for mobility is literally omnipresent, finding expression in ever so many ways. Let us consider designations, for a start. Designation is an important identification mark, an acronym summing up the bearer's position in the organizational pyramid and a commentary on his worth in the firm and beyond. Status-giving titles are therefore greatly sought after while pejorative ones meet with disapproval. In industry's lexicon, even generic labels have a hierarchical significance: worker (and its Hindi equivalent *mazdoor*) is the lowest form followed in that order by operator/technician, officer, manager and executive.

Bland labels such as 'operator', 'technician' and 'craftsman' are increasingly the mean, or minimum expectation, especially in high-technology industry where the more conventional 'worker' is itself an opprobrium. There are in any case few workers in the strict sense of the term, because hardly any one receives a 'wage' calculated as a daily rate. Almost all of the public sector calculates compensation by the month, and the private sector is quickly falling in line with this silent revolution of replacing wages with salaries. 'Wage' has in fact given place to 'salary' in the common parlance of industry. The addition of a status-giving prefix (senior/head/special-grade/master) can elevate a bland designation whereas the more odious prefixes (unskilled/semi-skilled) detract from its worth. At the upper reaches of skill, the aspiration is to break the class-barrier and escape all association with manual work by becoming a 'supervisor' (although not a 'foreman') or, better still, 'junior engineer', even though in the short run such upgradation may not result in monetary gain and perhaps even cause a loss.[5]

Industry is replete with evidence of the search for designations, especially expressive ones which stress rank and the progression from physical to mental work, and the creation of new layers of the hierarchy to accommodate the new designations. Thus, the head clerk of a multinational firm returns home one day to find his wife admonish their son to catch up with his homework if he did not want to end up like his father—a clerk. The next morning the distraught head clerk accepts a promotion as a junior officer which he had long declined because it would terminate his career as a union official. Head clerks in this firm take home a tidy packet—a better salary in fact than junior officers because of overtime—but the label eventually turned out to be more important than money. There is no shortage of clerks in industry who want to become 'officers', or at least 'office assistants', because that would help them find better grooms for their daughters. Employers who are well aware of these predilections have often used designations as a method of union-busting. Since workers can scarcely resist being called supervisors (nor, for that matter, can head clerks and store-keepers resist the opportunity to become junior officers), entire categories of employees have been upgraded to non-union positions to break the union's back.[6]

[5] Becoming a junior engineer is especially difficult, because management would normally not confer the title without an engineering degree. As a result, senior operators who have decades of experience at an instrument console in a fertilizer plant or a petroleum refinery, and therefore know the job inside out, make it a point to put in place freshly-arrived junior engineers who know what the books say but not what goes on behind the console. Conflict between the top rung of the blue collar and the first line of management is a routine occurrence. In contrast, clerical and secretarial staff find it easier to cross the barrier and become 'junior officers', apparently because specialist qualifications are not necessary as in the case of engineers. Once the designation of 'officer' is gained, becoming 'manager' is a short step which will come with the passage of time. Staff functions such as personnel and administration are the most common parking slot for typists, stenographers and secretaries promoted to managerial ranks.

[6] Any employee can join a trade union, but only workers, defined as those performing manual, clerical or technical work, enjoy the protection of the Industrial Disputes Act which is the fountainhead of job security. Supervisory personnel who lack this protective umbrella can be dismissed more easily.

Another manifestation of the search for mobility is the marked preference at all levels of industry for mental over physical work. Industry itself recognizes the superiority of the former over the latter, because even highly skilled physical labour is paid less than generalized and unskilled mental effort. While these differentials may be overturned in practice by uncontrolled overtime payments, upsetting them in precept is nearly unthinkable. Few organizations would, for example, consider signing a formal wage settlement which pays a highly skilled machinist better than a junior administrative officer who spends his time processing files, although the machinist possesses a scarce skill and does in fact gross more than the officer by virtue of overtime. In what seemed like an ingenious alternative to promotion policies which convert skilled workers into unskilled officers, an enterprise engaged in exploring for oil and natural gas re-christened a senior carpenter 'Assistant Engineer-Carpenter'. The expectation was that the first part of the new designation would satisfy his urge for mobility with a status-giving title and the salary that went with it, while the second part would keep him in skilled manual work where he was most needed. The carpenter had other ideas—he wanted to be like other assistant engineers, none of whom did any manual work.

A longitudinal view is perhaps a useful way of driving home the centrality of mobility to workers' aspirations. Three decades ago when the factory in question came up adjacent to a major industrial city, it recruited young boys and girls from working class families. Few had gone beyond school because their parents could not pay for their education. Some recruits were put on to familiar manual jobs, but because of the process technology employed by the factory, a good number went on to become either skilled technicians maintaining sophisticated machines or process operators presiding over instrument panels. Their earliest struggles were to gain parity with textile workers in the nearby city whose powerful unions had won them good wages and enviable bonuses. Thirty years later, they compare themselves not with textile workers who are thought to be illiterate, unskilled and lowly, but with chemical workers all over the country who are paid to exercise discretion and responsibility and not to flex muscle. Everyone wants to be transferred to the automated processes even at the cost of reduced pay, and the management is unable to assign manual jobs to any but freshly recruited apprentices. Workers have invested in substantial property—there is hardly any one from the earlier crop who does not own a house with its own compound and little garden.

Their children go to English-medium schools named after Christian saints which have mushroomed in the neighbourhood.

A skilled technician maintaining some of the most automated plants in the factory reminisced on his experiences spanning some twenty years. Barely forty years of age, he had reached the top of the blue collar grade. He dressed well, spoke fluent English, read technical journals to keep up with developments in his trade, and had a demeanour that was no different from that of a middle manager. He claimed to be so skilled in his job as to achieve substantial improvements in productivity with marginal additional investment, if only he had the power to make those decisions. There was in any case no question of any of his managers, all with engineering qualifications, being able to run the plant if skilled workers like him chose not to cooperate. With another twenty years to go before retirement, his lament was that he could not become a manager merely because he had not gone to college. He demanded to know if it was a fair system that allowed such a formality to hold him back even though he could match any of his managers when it came to technical expertise.

Designations are just as important to managers. Among the more significant developments in industry in recent times is the proliferation of levels and labels in the management cadre. Managers consider it burdensome to carry the same designation for any length of time, and quick promotions have become the common method of alleviating this sense of stagnation. Without an appropriate rank and label, companies would certainly find it difficult to retain their best talent. Caught between the demand for status-giving designations (as, for example, deputy general manager, general manager, senior general manager, group general manager, executive director, associate vice president, group vice president) and the limit to the creation of functions which go with these labels, companies have chosen to give a change of designation without a change of job. As a result, a promotion may get the recipient a superior title and the perquisites which go with it (an additional airconditioner, a better carpet, a bigger table) but neither significantly increased pay (thanks largely to the compression of differentials) nor the job content appropriate to the title. This is an especially common predicament in the public sector: one can find no fewer than a dozen general managers in some of its plants. The designation is nevertheless extremely important as a symbol of rank in the firm and a measure of worth in the wider society.

ORGANIZATIONAL STRUCTURE AND THE EXERCISE OF POWER

Several consequences follow as a result. Modern management theory suggests that flat organizations with the fewest possible levels are intrinsically more efficient, and instances have been reported of firms with no more than half-a-dozen levels from head to foot. While theory prescribes a delayering to create flatter structures, the Indian corporate sector is actually adding further layers. The typical organizational chart would show structures which are neither flat and modern, nor even traditionally pyramidal, but shaped like a barrel, with a bulging midriff and (counting out the chief executive) a flat top not very much smaller than the bottom. The divergence between theoretical prescription and actual development is just as acute with workers. It is now well recognized that multiple layers and descriptive designations are an important reason behind job rigidity and poor labour utilization. A descriptive designation tells not only what is expected of a worker, but equally what is not expected. A 'crane operator—mechanical shop' understands from his designation that he cannot be asked to work in the electrical shop. A 'senior technician' who thinks of himself as different from a mere technician without the prefix understands that he cannot be asked to substitute for the latter. Levels and designations created to meet social aspirations (although not only for that reason) are the fountainhead of restrictive practice.[7] The problem however is that while flat organizations may be superior, they cannot satisfy aspirations in a society where a person's sense of worth is determined by the number of levels below. The feasibility of flat and egalitarian structures in a hierarchical society shaping industry in its own image is debatable.

The system of accelerated promotions has major implications for power and its distribution. One of the conundrums of the corporate sector is that managers complain of insufficient power, even as they hesitate to exercise the powers they do have. Consider, for example, an assistant manager who goes on to become deputy manager, manager, and perhaps even assistant general manager by the time superannuation appears on the horizon. These attractive designations have been

[7] Another major reason for descriptive designation is the widespread managerial belief that without precise job descriptions it would be difficult to get any work out of labour. The designation thus becomes a cryptic resumé of the job.

robbed of much of their shine by the loss of functions and powers they were thought to have. The new incumbent is also troubled to find that promotion has not enlarged his sphere of control. Thanks to the uncontrolled growth in supervisory and managerial personnel, most organizations have in fact witnessed a steady contraction of span. Some especially obese enterprises in the public sector have, for instance, a staff-to-labour ratio of 1 to 5 or even less. In a society where status is judged by the number of subordinates, the decline in span comes as a jolt to aspirations. The response of managers to this situation is a sense of powerlessness, a feeling which grows especially acute as the fascination with titles wears off. There is much complaint that power is centralized—that it does not percolate sufficiently to lower levels to permit the satisfactory discharge of responsibilities. The typical response of every layer of managers until one reaches the top two or three rungs is that power vests not with themselves but with the level immediately above.[8]

In the midst of such widespread complaint that power is insufficiently delegated, there is also a marked reluctance to exercise the power that is actually available. Jobs do not remain entirely static even where promotions are quick and easy, if only because new entrants at the bottom would have little to do when every level carries its functions to the next rung. An accretion of functions and the powers that accompany them, even if modest, does therefore take place, especially with promotion to the upper echelons of middle management. The typical response of an entire array of middle managers to situations that call for decision-making is, however, to hide behind rules, search for precedents, pass the buck on to the next level, and generally avoid exercising the powers they do possess.[9] Entrepreneurial,

[8] In a large public sector plant with 10 levels among executives, every level came up with this response until assistant general managers had been reached. They admitted to having the power to take decisions, but there were only two levels above them—general manager and executive director.

[9] The case of a large public sector firm is illustrative. Their policy is to reward 'excellent' performance for three consecutive years with accelerated promotions and to deny promotion to those with a consistently 'poor' record, but this has never been implemented. There is not a single case of such reward or punishment for as long as anyone can remember because middle managers who evaluate workers' performance play safe by lumping everyone in the middle ranges of the scale. No one either succeeds or fails, as a result. Discussing the company's poor performance, a trade unionist remarked that they desperately needed someone to succeed. Managers argued however that quick promotions aimed at motivating a few would in fact demotivate the majority.

risk-taking behaviour is particularly hard to come by at any but the highest levels. There are contradictory explanations for such seeming abnegation by the very people who complain about powerlessness. Middle managers argue that they would rather play safe because the levels above would not stand by them if a decision miscarried. The most common explanation at the higher levels is, however, that promotion policies are to blame. Managers promoted from the ranks who bring a bureaucratic, rule-bound orientation to their jobs, claim senior executives, have a trained incapacity for risk-taking behaviour.[10]

Whatever the truth of these conflicting claims, the reluctance in the middle ranges of the managerial hierarchy to exercise power is very real. We have once again a situation where theory and practice are pulling in opposite directions. Decentralization of decision-making is at the very centre of recent developments in human resource management. A range of decisions relating to work organization is percolating not just to supervisors and first-line managers but beyond them to workers on the shop floor. The logic behind this development is that decisions become more acceptable and less expensive when pushed down the organizational ladder. In the Indian corporate sector, on the other hand, decisions have travelled up the ladder along with people. The cumulative effect of proliferating layers, time-bound promotions and the equivocation of an entire range of middle managers towards power and its exercise has certainly been to move decision-making up rather than down.

THE SEGMENTATION OF THE LABOUR MOVEMENT

Having considered in some detail the response of the work organization to social pressures, we ought to turn to the labour movement. In the early years of this century when trade unionism was in its infancy,

The view at the top is that middle managers choose the path of least resistance because reward and punishment might have to be defended before the body of workers.

[10] Among the most significant instances of this is a public sector firm engaged in international trade on the government's behalf. The company handles bulk imports of such commodities as sugar, edible oils and foodgrains, but some 60 per cent of managers up to the rank of deputy general manager are secretaries, typists, accountants and office assistants climbing up the career ladder who are unfamiliar with the intricacies of international trade.

labour's response to society appeared to consist essentially in its choice of leadership. Industrial workers who were drawn from the lower strata of society needed the evangelical assistance of upper caste politicians and professionals if they were to have any hope of dealing with capital on fair terms. Beyond this, trade unions reflected political ideologies more than social cleavages. Indeed, they seemed to be especially adept at breaking primordial barriers and levelling status distinctions. In one of the few systematic attempts at examining this facet of unionism, Uma Ramaswamy (1979) argued that political convictions were a major force reducing social distance between upper caste trade unionists and their Scheduled Caste colleagues in the Coimbatore textile industry. Some years earlier, it had been argued in similar vein that there was little evidence of trade unions in this industry being organized along ethnic lines (Ramaswamy 1976). What has happened to trade unionism in the years since these data, limited as they were to one industry in one corner of the country, were collected?

The most noteworthy development of the last two decades is the segmentation of the labour movement on an axis different from, and increasingly more significant than, the ideological cleavages of the past. While mainline trade unionism clings to the political divisions inherited from the early years of this century, the more important fact for a growing section of the labour movement is the new segmentation reflecting more closely the forces of the wider society, not necessarily primordial. The segmentation is along the lines of social class and occupational status: middle managers and supervisory staff occupy one end of the labour movement with their 'officers' associations', unions of underprivileged impermanent labour bring up the other end with their own separate union, and the space in between is occupied by permanent workers and white collar clerical staff with a sense of exclusiveness that sets them apart from both ends.

Middle managers and supervisory staff are a growing section of the labour movement, especially in the burgeoning public sector which employs them in vast numbers. They would not join a union of workers, nor even refer to their formations as trade unions, although they are in fact unions registered under the same law as any other. Officers' associations, as managers' collectivities are known, reflect their members' position in the organizational pyramid and status in the wider society. They pursue trade union objectives but hesitate to use trade union methods. Paradoxical as it may sound, one of their major demands is the recognition of managers as 'workers'. Since only workers (as defined in note 6) enjoy the employment security guaranteed by the Industrial Disputes Act, such recognition has become the main plank of the National Confederation of Officers' Associations, their apex body. Other demands are often union-like: collective bargaining rights, quicker promotions, periodic wage revisions, better neutralization of living costs through dearness allowances and so on. Managers are reluctant militants who will resort to union-like behaviour only if they have to, preferring instead the gentler method of interest representation. Associations are often stand-offish towards unions (even though some of their leaders are union activists promoted from the ranks): the two may cooperate in a crisis but will rarely work towards a common agenda, much less a common identity. With all this, one of the commonest accusations by higher levels is that managers are behaving like trade unionists.

At the other end, contract and casual workers have a separate union not because they want to but because the permanent workers' union will not take them on. They are the underclass of the corporate sector, doing much the same work as permanent labour (often very much more) and paid a fraction for it. They have just one demand—confirmation as permanent hands. Since this is nearly impossible to achieve, the next best is to ask for parity with regard to wages and benefits. Permanent workers do not ignore this underclass altogether, but would not make common cause with them either. The more typical response is to extend moral support from a distance or engage in an occasional bout of sympathetic action. Taking a less charitable view, it suits permanent workers to have an underclass that will take on dirty or hazardous jobs, and be thankful for small mercies.[11]

That takes us to permanent workers, the true occupants of Holmstrom's citadel, with better pay than the harder working impermanent labour and better security than the more qualified managers. Their interests have changed visibly in recent times. While their unions continue to demand stiff wage increases and some more money is always welcome, they are likely to identify mobility rather than compensation as their prime concern. Invidious distinctions emphasising their subordinate status with respect to managers trouble them enormously. Throughout the country one can find unions demanding to know why managers need exclusive dining rooms, exclusive transport, separate

[11] I have witnessed permanent workers ask contract labourers to take on some of their work in return for a canteen coupon or other similar favours.

entry and exit points, and even separate toilets. The union apart, new management theories inspired by Japanese work practices offer powerful ideological support for egalitarianism. Open offices, shared canteens and common systems for punching in attendance are becoming the norm, especially in greenfield sites where there is no privileged group to resist the loss of privilege. Enterprises driven more explicitly by Japanese management styles may go beyond shared facilities to promote common uniforms, calisthenics and such like.

None of this is of course easy in older, brownfield plants where executives are apt to see the change not as a move towards egalitarian values but capitulation to union pressure. They are right in some ways, because permanent workers are not engaged in an ideological crusade for the spread of egalitarianism: they are merely interested in a better deal for themselves. Their quest is for equality with those above in the ladder, not with those below. Therefore, workers who accuse managers of being elitist have no qualms about refusing to eat the same food as contract labourers. Having converted their enormous bargaining power into better wages, workers and their unions are now trying to convert it into better status. Egalitarianism is an incidental accompaniment to this quest.[12]

Mobility up the ladder, if possible into supervisory and managerial positions, is another important objective that we have already alluded to. Although trade unions have lent the full force of their bargaining power to extract favourable promotion policies, workers must still equip themselves with better educational qualifications to make the most of the opportunity. University degrees, especially in the liberal arts and social sciences, gained often through correspondence courses, are becoming common among a labour force which was until recently considered to be illiterate, ignorant and forced into industrial employment against its will. Table 1.1 lists the educational qualifications of non-executive employees in a large public sector plant manufacturing electronic equipment and components, conveying an idea of the employee profile in large scale industry.

[12] In an interesting recent example, highly skilled technicians and operators in a process plant went on strike demanding promotion to supervisory positions. With the management insisting that promotion across the class divide required university education, the strike continued for some four months. At the end, they settled for a better designation and a uniform in recognition of their superior status. They are now called master craftsmen and master operators, and wear uniforms which set them apart from the rest of the permanent workforce.

Table 1.1 ***Educational Qualifications of Non-Executive Employees in a Public Sector Plant***

Education	*Men*	*Women*	*Total*
No formal education	491	51	542
Between 4–9 years of school	1,854	40	1,894
Completed 10 years of school	1,214	1,485	2,699
Diploma in typing and shorthand	440	436	876
Diploma from industrial trg. inst.	2,286	43	2,329
Completed 2 years of college	141	126	267
Diploma in engineering	448	101	549
Diploma in other disciplines	153	28	181
Graduate in science/hum./commerce	387	119	506
Graduate in engineering	17	0	17
Post-graduate and above	69	9	78
Other qualifications	9	3	12
Total	7,509	2,441	9,950

Note: The data are for the year 1992. Being 40 years old, the plant has an ageing workforce—only 2 per cent of the employees are below 30 years, while 61 per cent are between 40 and 50 years, 19 per cent are above 50 years. Workers of younger firms in electronics and similar other sunrise industries will certainly have better qualifications. On the other hand, employees of chimney-stack industries (such as textiles, jute and coal) will probably be much less qualified.

Source: Data collected by the author.

At least some of the 600 workers and white collar staff in Table 1.1 who have a university degree will make it into supervisory and managerial positions. While a few might rise to become middle managers before superannuation, others would strive hard to shed such unattractive designations as foreman, first-line supervisor and head clerk and become junior engineers and junior officers. Promotion and designation apart, educational qualifications bring more immediate monetary rewards. Numerous unions have bargained for an additional increment to reward workers who collect a degree or diploma, no matter how unconnected with their actual job. There is growing acknowledgement among human resource managers that workers are likely to become demotivated unless industry recognizes these aspirations and creates avenues for their fulfilment.[13]

[13] The case of a watch factory in south India is indicative of the initiatives that might be expected of employers. Managers visited every secondary school in the rural districts of the state to administer aptitude tests to the pupils, probe into their parentage,

Trade union power also translates into jobs for children of permanent workers, a point I have already touched upon. Apart from reducing the quality of labour (see note 3), this has consequences for labour-management relations which can be scarcely imagined. With the job literally becoming part of the patrimony bequeathed to the next generation, workers use union power to stall decisions that might jeopardize their children's interests. Automation and re-deployment of labour are opposed because they can downsize employment and evaporate the patrimony. The confirmation of temporary hands does not find favour because it cuts into the job opportunities for their own children. Attempts to step up machine speeds and revise workloads cause workers to wonder if their children can cope with the increased tempo. They want educational qualifications for recruitment lowered to accommodate wards who could not clear school, and automatic promotions to guarantee their mobility. Recruitment is such an emotive issue as to pose a threat even to the union. Because they can be accused of complicity with the management to syphon scarce jobs to favoured candidates, union leaders want employment to be distributed on the basis of the parents' seniority and not any selection procedure that might smack of subjectivity. While this might rob the firm of the little discretion it had in the matter of recruitment and cause a further decline in its human resource quality, the leaders' own concern is to avoid internal dissensions which can weaken the union and even break it up. Most unions are under pressure from members to demand preferential recruitment of their progeny, but few have the power to go back on it even if they realize the dangers that lie ahead for themselves and the firm.

The social aspirations of the citadel's inhabitants have been steadily rising for the past decade or so, and the future will in all probability witness a further reinforcement of these ambitions. Knowing that it is the bargaining power of the trade union which enables workers to convert money and job security into social status for themselves and their children, managers too might come to view collectivization as a

---

verify their antecedents, and assemble a 2,500 strong workforce consisting of the brightest boys and girls. Those who displayed particularly high levels of skill and concentration were trained for maintenance jobs while the rest were sent to the production line. Recognizing that such intelligent workers would become resentful and demotivated if denied the opportunity to move up, the company is hoping to set up a technical institute to impart higher education.

method of enhancing their own status, or at least of defending the steadily eroding differentials in pay and perquisites between themselves and workers.

Our description of the segmentation of the labour movement would be incomplete without mention of yet another facet, but one whose roots lie in the traditional cleavages of Indian society—the SC/ST (Scheduled Caste and Scheduled Tribe) association. Almost unknown a couple of decades ago, these associations now have a ubiquitous presence in industry, especially in the public sector. Not being a substitute for trade unions and officers' associations, they leave aside general economic demands and concentrate on issues specific to these social categories, such as preferential recruitment and promotion. There is much heartburn among the rest of the workforce over affirmative action, and in particular over promotion policies which enable members of these categories to move up the ladder faster than the rest. A consequence of this heartburn is the improbably named non-SC/ST associations. Aimed at bringing together the rest of the workers to fight against special privileges, their activities ebb and tide in tandem with those of SC/ST associations.

THE FUTURE

While it may be something of a truism to say that industry reflects the values of society, the interplay between social values and the principles of industrial organization is as fascinating a field of study as it is neglected. The values which have been working their way into industry have changed vastly since the early years of this century when the Royal Commission on Labour reported that the pull of village society was so strong as to make it difficult for industry to recruit and retain a labour force. Although later research did much to nail the theory that labour was uncommitted to industrial work, the emphasis for the next several decades in such research as cared to probe this issue was on the extension of primordial links and loyalties to industry. Factory hierarchies thus seemed to correspond with the social hierarchy outside, with the upper castes supplying professional and white collar employees, the peasant castes blue collar workers, and the untouchable castes menial labour. There were other continuities as well—Muslims took to weaving in the textile mills which others would not take up because it involved sucking the yarn through the shuttle and therefore contact with spittle; leather workers took to tanning; and workers of different castes drank water from different taps and ate their food separately. Indian society seemed coterminous with its structural division into ranked social groups, each tied to a specific occupation, and it was this structural division rather than any underpinning ideology, ethos or value system which seemed to find reflection in industry. Perhaps it was the proclivities of the researchers which led them to look in industry for the mirror image of that distinguishing feature of Indian society—the caste system. Perhaps industry itself was too nascent and unsophisticated to afford opportunities for anything else.

With growing size, stature and technological sophistication, industry has become an important avenue not so much for replicating the structure of the wider society as giving expression to its deep-seated values and beliefs. Although the two are admittedly difficult to separate, it is clear that attempts to superimpose the social hierarchy on the industrial hierarchy would yield little result today, whereas ample evidence can be found of the inroads made by social values. Because it pays so well and offers careers and not mere jobs, industry is able to draw into its vortex every segment of society even if the position on offer is not in accord with its location in the social hierarchy. Fundamental social values are however sweeping industry as never before.

The most pervasive social process being witnessed in industry is the conversion of wealth and power into status. If, in the traditional society, the search for status led upwardly mobile groups to employ bards and genealogists to write caste myths and imitate the social mores of superior castes—a process Srinivas termed sanskritization—in industry it involves the pursuit of values and symbols which are acknowledged to be status-giving. The preference for mental over manual work, compensation systems and promotion policies which legitimize such preference, the creation of new organizational layers and high-sounding designations, and attempts to dismantle others' privileges while defending one's own are all manifestations of this process. For an increasing number at all levels, the upper end of the search for status is to launch one's own firm or business. The labour movement reflects these forces as much as industry itself. Although workers have some shared interests with middle managers above and contract labourers below, and could combine with either, or perhaps even both, to create a formidable force, status aspirations come in the way. Only contract labourers who have everything to gain, and nothing at all to lose, would unequivocally support such a move. Workers would like to join forces with middle managers and not contract labourers, while middle managers would team up with no one although an alliance with workers would do wonders to their bargaining power.

Indian experience with industrialization is in many ways distinctive, reflecting as it does the values of the society and the aspirations of its people. The social forces we have considered must of course not be viewed deterministically since industry can influence society as much as it is influenced by it. Moreover, other forces and pressures, equally germane to our society but different from the ones now visible, could bring themselves to bear on industry with the passage of time. Since industry can nevertheless not be cocooned from society, culturally unique ways of organizing industrial activity seem inescapable. The question is not whether there are universally valid organizing principles—there admittedly are—but whether entire models can be transplanted across cultures. Moreover, the universal and the culturally unique are not exactly easy to separate in any model, whether it is the Anglo-Saxon propagated with such enthusiasm by Kerr in the early sixties but now stands discarded, or the Japanese which currently holds sway. Some aspects of Japanese industrial organization, for example, have been argued to be not universal even to Japan but specific to a corporation—Just-in-Time, one of the most powerful tools in the Japanese repertoire, has thus been claimed to be specific to Toyota (see Wood 1991). Attempts at wholesale transplantation of the Japanese model in other milieux have often run into stiff cultural resistance.[14]

The point, moreover, is not whether there are components in any model which have universal application and can therefore be transplanted in a different social context. The more salient question is whether there are cultural attributes in a given society which can be harnessed to the advantage of industry. It stands to logic that people do well what comes naturally to them. As Indian industry shifts from a regime of controls, with its protected markets and guaranteed profits, to a competitive environment in which the fittest survive, these questions will come up forcefully. With human resource productivity, universally acknowledged to be the key to successful industrialization,

[14] See, for example, Fucini and Suzi (1990). The book is a shop-floor account of such resistance in a Japanese automobile plant in the US.

moving to the centre of stage, we will be forced to ask what values there are in our society and culture which can help industry to get more out of its people. Right now, the answer to that question can only be that we do not know. Not only has it not been necessary for a protected industry to ask that question, but the touchstone for evaluating the impact of such social forces as I have considered is itself a set of organizing principles evolved in another context.

REFERENCES

DORE, RONALD. 1973. *British Factory, Japanese Factory: The Origins of National Diversity in Industrial Relations*. London: Allen & Unwin

——. 1989. 'Where Are We Now? The Musings of an Evolutionist', *Work, Employment and Society*, 3: 425–46.

EDGREN, GUS (Ed.). 1989. *Restructuring Employment and Industrial Relations*. New Delhi: ILO.

FUCINI, JOSEPH and SUZI FUCINI. 1990. *Working for the Japanese*. Glencoe: Free Press.

HOLMSTROM, M. 1976. *South Indian Factory Workers: Their Life and their World*. Cambridge: Cambridge University Press.

KERR, CLARK, J. DUNLOP, F. HARBISON and C. A. MYERS. 1973. *Industrialism and Industrial Man*. Harmondsworth: Penguin.

MATHUR, A. N. 1989. 'The Effects of Legal and Contractual Regulations on Employment in Indian Industry', In Gus Edgren (ed.), *Restructuring Employment and Industrial Relations*. New Delhi: ILO, pp. 153–202.

RAMASWAMY, E. A. 1976. 'Trade Unions and Caste in South India', *Modern Asian Studies*, 10: 361–73.

RAMASWAMY, UMA. 1979. 'Tradition and Change among Industrial Workers', *Economic and Political Weekly*, 14: 367–76.

——. 1983. Work, Union and Community: *Industrial Man in South India*. Delhi: Oxford University Press.

WOOD, STEPHEN J. 1991. 'Japanization and/or Toyotaism?' *Work, Employment and Society*, 5: 567–600.

2

Caste and Status in a World of Technology

WILLIAM H. NEWELL

In her book *Work, Union and Community: Industrial Man in South India* (1983), Uma Ramaswamy discusses caste in Tyagipuram, a suburb of Coimbatore consisting almost entirely of mill workers. According to her:

> Caste is not a major principle of social organization in Tyagipuram. When people speak of caste, the referrent is the diffuse clusters which share a similar ritual status. ... Even as a cluster, caste has little place in the daily routine of the worker. By contrast, the worker is in close and constant contact with his colleagues and union members (1983: 118).

Using one particular factory in Kanpur in Uttar Pradesh as an example, I shall demonstrate that the use of the term caste in an industrial

Note: This paper was written over thirteen years ago. The Lal Imli factory, in this form of covering the whole production process from raw wool to finished cloth, is now closed. The economic reasons for total change can be seen from this paper to be based on a need for structural organization. A factory is *not* a society.

situation can be better understood if we look at it as referring exclusively to status. Within a factory, status originates from one's task within the factory; within a city, status originates both from the task one undertakes within the factory and from the importance of the particular firm to which one is attached; within the all-inclusive urban situation, status originates from a multitude of different contexts based on power (politics), religion (ascribed connection to the *varna* system), or personal wealth. I shall confine this paper only to the position of the worker within the factory. Very often, when an Indian worker refers to caste, he only means status inside the factory.

The city of Kanpur in northern India, about 80 miles south of Lucknow on the river Ganga, now has a population of over a million, making it the largest city in Uttar Pradesh. It is surrounded by a very depressed rural population which receives special aid from the state in the way of agricultural subsidies. For all practical purposes it had no urban or industrial history prior to the mutiny of 1857 when it consisted of only a few small villages. Its early industrial expansion was based on a military market for a cantonment which was established just outside these early villages in an area which was violently anti-British.[1]

The military required mostly boots, saddles and clothing. In 1863 the Government Harness and Saddlery Factory was founded, and in 1864 a cotton mill was established called the Elgin Cotton Spinning and Weaving Co. Ltd. This mill went bankrupt shortly afterwards, but was the precursor of a later Elgin Mill founded in 1880. A number of tanning works were also established in this early period. The capital for most of these early large works came from overseas and the entrepreneur managers were white. One interesting feature about these early large factories with a labour force of round about 5,000 is that each new factory was founded by an industrialist from a previously established factory from the same area—something like the Japanese *honke-bunke* system. Later foundations regarded themselves as junior to the earlier foundations, especially in respect to the mills. If one asks the present day textile workers to grade factories in terms of status in respect to working conditions, good management, and so on, the hierarchical order is approximately the same as the order of founding, with the earliest founding firms being regarded as better to

[1] In 1852 the cantonment occupied 25.92 sq.km, the civil lines 14.23 sq.km, and the city proper only 2.72 sq.km. In 1847 the city population was 85,821 (Singh 1972).

work for. Cotton mills continued to be founded up to 1911 with the establishment of the Swadeshi Cotton Mills (with Indian capital). Singh (1972) regards the first phase of industrial growth to be from 1857 to 1914.

The second phase of industrial development was the period between 1915 and 1947. Large factories were set up to manufacture fertilizer and heavy engineering goods, especially railway wagons, and were later replaced by ordnance factories. New factories were built mostly under the stimulus of the two wars, but practically no increase in numbers or size was made between the 1920s and 1930s. In the early thirties there was an advance with new factories being set up under the J.K. group of companies for jute, cotton, sugar and light engineering, amongst others—but the style of functioning was different. The J.K. companies were under the control of a family-managed agency. In Kanpur this was marked by a large administrative tower which dominated the centre of the city. Whereas in the early period the interest of the owners of the factories seemed to be in the technical organization of the factories themselves, and many of the managers had technical qualifications in textile or other forms of engineering, this later period seemed to emphasize factories mostly as a form of investment exclusively, with salaried technical managers who did not remain in the factories all their lives but moved within J.K. or left. The J.K. firms in Kanpur all came under one central authority irrespective of what they manufactured. The J.K. factories in Bombay and Calcutta were separately controlled by other members of the family.[2]

[2] Although I was not able to visit a sufficient number of J.K. factories to express a firm opinion, I did look at one particular small factory, J. K. Satoh, making small mechanical hand ploughs, originally under Japanese licence. The real purpose of this company was to obtain special government subsidies for new industries and, in my opinion, not to make a profit from sales. In one particular year they sold only half a dozen ploughs. There is a fundamental ideological difference between a manufacturing entrepreneur and a commercial entrepreneur. I consider this company to be a commercial entrepreneur whose aim was to extract a profit from government subsidies. While I was in Kanpur an important J. K. family marriage took place and almost the entire expenses of this marriage were borne by one particular J. K. company, rather than from the income of the family—an action of a commercial rather than an industrial entrepreneur.

This period is also marked by the rise of two manufacturers' chambers of commerce. Although both were political, there were differences in respect to membership and the nature of the ownership and use of capital by the members. Members of the newer chamber were all Indians with Indian capital.

The third phase of industrial development is from 1948 onwards. This period is marked by a great increase of small factories in light industry and specialist manufacturing, including artificial fertilizers. Motor repair shops and speciality food shops have also become very important.

The history of industrial development in Kanpur is similar to a geological strata. At the bottom are large factories of cotton and leather, employing mostly unskilled or semi-skilled labour. In 1962 the smallest number of persons employed in a single textile mill was 2,410, and the largest number in a single mill was 6,909 in Elgin Mill No. 1. The largest leather boot factory employed about 1,200. Practically every one of these factories was founded in the first phase. While these factories continued to function, new jute, ordnance and other factories made the second wave. Then a third wave of medium engineering and more specialized factories developed. In each wave the factory unit became smaller and more specialized, and required a higher proportion of skilled workers or trade specialists. In 1963 the largest metal and engineering works (excluding the government ordnance factory) was the India Rolling Mills with about 300 employees, and the smallest was the Safe and Wheel Industries making rickshaw parts with 15 employees. Of the 32 registered iron and steel factories at that time, the average number of employees was about 80. What is especially interesting about the development of industry in Kanpur is that once one layer had been replaced by another, no new firms of an earlier type were built. Some of the first type (namely, cotton and woollen mills) are now facing problems, and practically all the mills in Kanpur, without exception, are either very sick or have been taken over by the government. But it should not be forgotten that it is the older mills in Kanpur which were often the last to surrender. The Lal Imli mills (to which I will turn later) held its centennial in 1976. Since this paper has been written, it has now been totally taken over by the government.

The proportion of skilled to unskilled workers is constantly rising in the newer (and more profitable) industries in Kanpur. At the same time the number of registered factories is growing faster than the number of workers employed in those factories. Whereas in 1921 there were 38 registered factories employing 20,706 workers, in 1948 there were 179 factories employing 89,289 workers, and in 1963, 356 factories employing 66,278 workers. Although I do not have later figures, I know that while in 1978 the number of factories had increased to 442, the number of employed workers had hardly increased over the earlier period in registered factories. It is clear that the pattern of factory work has radically changed over the last hundred years in Kanpur: namely, smaller factories, more skilled workers in registered factories, and a lowering of the proportion of people employed in the various firms.

Table 2.1 shows the number of factories of various types and the workers employed in them in 1962. Since 1972 those in textiles and leather have decreased markedly with a number of mills being closed.

Table 2.1 *Types of Factories and Workers in Kanpur in 1962*

Type	*Factories*	*Workers*	
		No.	*%*
Textiles			
(a) Mills	15	63,000	59.2
(b) Spining and ginning	22		
Leather	37	6,638	6.0
Metal and engineering works	34	5,298	5.0
Food processing	54	3,114	3.0
Chemical and plastic works	47	2,050	2.6
Printing and publishing	19	625	0.6
Wood and cardboard works	11	286	0.3
Electrical accessories	12	552	0.5
Miscellaneous	107	24,772	23.4
Total	358	106,335	100.0

Source: Singh (1972: 88).

It is well known that the proportion of low castes in the older cities of India is much higher than in the newer cities with more light and recently-founded industry.[3] In Kanpur, as industry becomes able to employ a higher proportion of skilled workers, the opportunities for unskilled workers to find employment become less. In tanning factories, the process of manufacture consists mostly of applying various chemicals to the leather, often by hand, and in such factories the majority of workers are unskilled and of low caste. In the rural areas immediately around Kanpur, the difficulty of finding work in the

[3] D'Souza (1975) shows how newer cities in Punjab have fewer residents from Scheduled Castes proportionately, because of fewer opportunities for unskilled workers in new industries.

countryside applies equally to all castes. However, as far as the low castes are concerned, as soon as they reach a certain age, irrespective of the degree of education they may have received, they try to obtain semi-permanent jobs in such industries as tanning or else desperately hire themselves out in the 6.30 a.m. free labour market in the middle of the city at (in my opinion) almost starvation wages. On the other hand, higher rural castes under no circumstances allow their sons to enter the urban labour market until they are at least fully literate, so that when the new types of jobs are created, the low castes are ineligible due to lack of literacy and training. Thus, as most jobs become available in the more newly established industries, of which only a small proportion are unskilled, the children of the unskilled workers in the older labour-oriented industries have fewer opportunities available unless they become educated. Modernization has little hope for increased lower-caste employment. Factories are not at all interested in the caste of the applicant where there are plenty of jobs available, but are concerned only with the job being carried out successfully for as little monetary expenditure as possible. This encourages educationally qualified candidates to only apply for the increasing number of skilled jobs. Education acts as an important sorting house for admission to the workforce. Why low castes from the rural countryside around Kanpur do not achieve a higher level of education cannot be ascribed only to a lower income, as even low castes with permanent jobs do not seem to see at all clearly that to be a skilled worker requires education to a higher level than that of a labourer. The situation in Kanpur seems to be somewhat different from that of Uma Ramaswamy's study of Tyagipuram.[4]

The external industrial situation in Kanpur rests on the different opportunities for different classes of workers in different firms as far as social mobility is concerned. Apart from government firms (such as ordnance factories) which require separate study, the highest skilled

[4] There are a number of other cultural features which distinguish Uttar Pradesh from Tamil Nadu as far as work is concerned. In a factory making torch cases in Lucknow, no women were employed on the assembly line although, in some other parts of India, a similar process required women. The manager claimed that the introduction of one woman would precipitate a strike as it would be regarded as taking bread from a family head. One other factory making electrical products run by J.K. and employing women from the beginning went bankrupt. I suspect that this was partly because the factory controllers looked upon the project as a commercial rather than an industrial one. The employment of women was blamed for its demise.

workers, namely the weavers,[5] saw no opportunities of moving elsewhere from the firms they were in, and a lot of their energies as workers were devoted to retaining their jobs in that particular firm. I will refer to this again shortly.

However, in the medium to light industrial firms making specialized objects, among the skilled fitters and turners there was a constant movement from low status firms (like J. K. Satoh) to high status firms where their opportunities were greater. In interviews with workers in the more specialized firms, every skilled worker who I spoke to stated that if better opportunities presented themselves they would move. Nearly all had entered higher status firms from lower status ones. Light engineering firms could easily be graded, with skilled workers always moving in one direction—up. These specialized firms (which often had over 80 per cent skilled workers) tried to retain their workers with long-service bonuses and retirement funds. One manager whom I interviewed stated that welfare was good business, as a skilled worker did not really earn his keep until he had been employed for six months. Thus it was uneconomical for such workers to leave as they would lose the employer's pension contribution. On the other hand, unskilled workers in skilled firms who could not be promoted due to their lack of skill often remained in the same firm all their lives. There is no managerial need for welfare funds for such workers as resignations are easily replaced. In firms in which the majority of workers were unskilled, they continuously battled with the management to try to become permanent. One particular firm which had a Muslim management automatically sacked their workers every 200 days, employed totally new unskilled workers, and then re-employed the former after another 200 days in alternate years. Horizontal mobility was in a sense forced mobility. Thus the difference between skilled and unskilled workers was not just a matter of a difference in education and salary but also implied a rejection by the company to consider unskilled workers as 'lifetime employees'. In Kanpur, unions for the most part did not accept temporary workers as members, and being a member of a union also implied being within the 'citadel', to use Holmstrom's phrase.[6]

This cleavage between different types of workers can clearly be traced back to the type of work and the nature of the workplace, not

[5] See Pandey (1990, Chapter 3) for information about the ideological attitude of weavers.

[6] Also see a recantation of the 'citadel' theory in Holmstrom's paper (1981).

caste or origin. Permanent and especially skilled workers emphasized education for their children to attain the same status, often lived together in the same part of the city when they came from the same firm, and used the unions they belonged to to emphasize their group status. For instance, in one textile firm the management granted an increase in the house rent allowance to all permanent workers. Within three days a group of foremen approached the personnel manager for a special larger increase in the housing allowance for foremen on the ground that their expenses were higher because they had to live in better houses and had to entertain other workers in their shifts to keep on good relations with them. Therefore, an increase for everyone at the same level was unreasonable. Naturally, a union of foremen would be created if their demands were not met.

In contrast, temporary unskilled workers fought with their employees on two grounds only—an individual desire to become permanent, and an increase in wages when the cost of living increased. Since they were usually unorganized and were not admitted to union membership, such strikes as did break out were usually quelled by depriving such workers of a portion of their wages or by force with the cooperation of the police. Unions only defended the interests of the permanent workers for the most part.

The situation of different workers clearly originates from the different working conditions and the work that is performed. The nexus between the work performed, the differential struggle for status, and the powerlessness of the unskilled worker is clearly more important than such cultural categories as caste.

Let us now examine status in one particular factory to illustrate the problems that arise when the work system does not coincide with the rank system based on the actual work performed.

The Cawnpore Woollen Factory (also called Lal Imli) had just celebrated its first hundred years of operation in 1976, when I visited it in 1980. It was controlled by a semi-government corporation—the British India Corporation—which owned the majority of the shares. Originally the factory was totally private, but the BIC owned a number of other companies in Uttar Pradesh and was responsible for appointing the manager and receiving the annual reports of the company. Altogether there were about 4,500 employees divided up into different departments arranged longitudinally along the production process. At the beginning was the raw wool warehousing department followed by the cleaning section; at the end was a sales warehouse for the finished product; and in between were various sections such as the No. 1 and No. 2 worsted departments, and the finishing department. Each department was headed by a graduate professional, and the departments were physically separated from each other. There were also one or two general departments such as personnel, or accounts and sales.

Certain departments (such as the boiler and stoking departments) had a high proportion of unskilled workers carrying out purely physical tasks. These workers all came from medium or low castes, not because the company had any policy towards appointing certain castes to certain positions but because only low castes applied for such positions. Other departments had a high proportion of specialized duties. Perhaps the most skilled was the No. 2 worsted department, in which out of a total complement of 133, 87 were twisters, spinners or winders. Of the remainder, not more than a dozen were doing totally labour-oriented duties, such as carrying the bobbins or finished cloth around. Prior to Partition, the majority of employees in this section were Muslims, but even now nearly all the weavers are either Muslims or from Hindu weaver or associated castes. The setting up of the machines, the arrangement of the warps and wefts, the constant changes in the pattern of the worsteds, and so on, are not tasks easily transferable to unskilled persons. Once appointed to a task, the weaver was not transferred to other tasks and received no change in remuneration except within fairly small limits. One could perhaps be promoted to the position of weaver *mistri* when one received more pay and undertook the same job or, exceptionally, became a foreman. But these vacancies were strictly limited. Each weaver also had a temporary weaver below him who worked only when the permanent weaver was ill or on holiday. Under the law, any temporary worker who becomes employed for more than 250 days should become permanent. Naturally the temporaries were not employed for longer than that period unless the full time worker had vacated his job. However, the temporaries were appointed in practice by the weavers themselves whose recommendations were always accepted by the management and who naturally chose persons from their own families or relatives. Persons were never transferred from one department to another, and there was no internal mobility between departments.

In the case of highly specialized jobs, it is perhaps reasonable not to expect job circulation, but over the whole mill there were about 320 job categories. Let me give an example from the finishing section. After the worsted is finally completed it is checked for blemishes on a huge roll, and one person with a sharp knife (called a cutter) cuts off the threads while another person with a pencil blots out such white spots which remain which have not been reached by the dyes. Any one after a few minutes should be able to do these tasks. Yet both cutter and penciller are classified job categories receiving a wage determined by the job. Before the war these classifications were carefully thought out so that one received an appropriate wage for the skills but, with inflation, workers receive a dearness allowance which is usually about twice the amount of the basic wage so that differentiation of wages by job distinction has been largely eroded away.

Indeed, wages and jobs are not the only form of status differentiation. Other criteria are the section of the mill in which one works, clothing, the type of language used with persons of the same or different status (usually a matter of intonation), seniority within the department, and membership of a trade union. Moreover, the truly skilled labour in the mill was very often of at least three generations' depth in Kanpur, had no regular connection with rural relatives, and owned houses. Those I visited were usually in special 'slum areas', but the houses were clean and often built by the family concerned so that it was their own property. These slums were very different from the *ahatas* (slum back housing), which were often occupied by unskilled labour, rented, and inhabited by groups of single men. (The distinction was very similar to that described by Rex and Moore in their study of Sparkbrook in Birmingham (1967). It is clear that the description of labour often presented by sociologists of labour, as being rurally oriented and constantly returning to their rural areas associated with a lack of labour discipline, does not apply to the skilled labour in the mill at all. The distinction is not between urban and rural labour but between skilled industrial labour and unskilled labour.

There is also a certain similarity here to a caste system in the sense that persons were occupationally differentiated, that positions in the system were passed down within the kin group, and that once one had acquired a suitable skill which could be used, the skills became differentiated into status systems and one could not move from one job to another. This relationship between skills was not however fixed so much by traditional rules as by the actual difficulty and skill of the task. On the other hand, there was no traditional caste differentiation. Persons ate together and joined the same factory organizations, and a person could not be appointed to a position unless he had first gone through the process of becoming a temporary worker which was, in part at least, a period of technical instruction in the duties expected of one.

Perhaps even more important, although certain jobs were dominated by persons of a certain caste or religion, the status position determined by the job was totally inflexible and one could not move up through the system by changing jobs. These jobs were fixed in number and were actually being slowly reduced to a similar level of skill by modernization. Thus, if one wished to improve one's status within the firm this could only be done by manipulating other forms of upward social mobility not directly connected with one's own job. I will take up only one example—that of trade unions.

The key date in the history of Kanpur unionism is the general strike which commenced on 2 May 1955, and was basically a demonstration against rationalization in the mills and a demand to receive reasonable wages after the war. It lasted 80 days. Among other objections, the doffers of the Cawnpore Cotton Mills 'refused to accept transfer orders from one department to another' (Pandey 1970: 103).

This strike was made possible by unity between six rival union federations—from the Congress union federations on the one side to the Communist and extreme right wing unions on the other. Out of 46,123 workers, 27,336 workers were directly involved. But this piece of successful working class unity marked the end of a phase rather than the beginning. From then on, both the number of union members and unions started to increase. For example, in 1956 there were 47,800 members divided into 117 unions, whereas in 1964 there were 91,500 members divided into 239 unions. In 1976 there were 453 unions registered by the Registrar of Trade Unions, although it is not possible to give the total number of union members as the figures are not reliable. But there is no doubt that the number of unions has increased much more rapidly than those unionized.[7] These unions were not all continuous. Between 1966 and 1971 there were 116 new unions registered and 53 deregistered. Between 1971 and 1976 there were 84 new unions registered and 28 deregistered. Of the unions registered in 1976 in Kanpur, their first registration is shown in Table 2.2. It is clear that unions die easily.

Any group of ten persons working in a firm may organize a union and have it registered with the Registrar of Trade Unions. They are

[7] In the whole of Uttar Pradesh, on 3 December 1976, there were 1,174 unions registered, of which 296 had a membership of below 50.

Table 2.2 ***Trade Unions in Kanpur by Year of First Registration***

Year of First Registration	*No. of Unions*	*Year of First Registration*	*No. of Unions*
1934	1	1960	3
1935	1	1961	13
1938	1	1962	15
1946	3	1963	10
1947	6	1964	9
1948	8	1965	12
1949	2	1966	10
1950	3	1967	15
1951	2	1968	9
1952	6	1969	24
1953	6	1970	19
1954	7	1971	26
1955	13	1972	12
1956	7	1973	20
1957	11	1974	26
1958	9	1975	7
1959	10	1976	14

Note: Tables 2.2, 2.3 and 2.4 may not always cross-tabulate as the figures were derived from two different sets of information. The figures were collected from the original registration applications.

recommended to use a standard constitution providing for compulsory auditing of accounts, and so on, and have an executive committee usually of seven persons. Over 50 per cent of the executive members must work in the firm. Although theoretically trade union members may belong to more than one enterprise, such a union based on more than one factory may not negotiate with employers under the UP Industrial Trades Dispute Act. Therefore, all working members belong to the one factory. No negotiations may be carried out until the union has been in existence for two years but a federation (which is a union of two or more unions) may negotiate on its behalf.[8] Table 2.3 gives the size and distribution of unions in 1976. Included in the 449 unions in the table are fourteen federations of unions, ten of which have political connections.

[8] The term 'union' in this article refers only to the group of members in one firm. The term 'union federation' refers to any large group to which unions may affiliate. I have collected all these figures from the Registrar of Trade Unions, Kanpur. The Registrar periodically and irregularly cancels the registration of all those trade unions which seem to have lost all their members.

Tables 2.2, 2.3 and 2.4 show that the proportion of affiliated unions is much lower for the smaller than for the larger unions, and the smaller unions are less likely to survive as corporate bodies than the larger ones. I have no figures to show whether those members of cancelled unions nevertheless remain unionized by belonging to other unions. The general analysis of these figures seems to show a situation roughly similar to the one in West Bengal (see Ghosh and Sengupta 1978).

According to the trade union law, for a union member to remain a voting member he should contribute at least Rs 3 a year to the funds of the union. In practice, however, a large number of workers do not seem to even contribute this amount. In the Cawnpore Woollen Factory, of the 4,300 odd workers, about 3,000 belonged to the largest union which traced its descent back to the Cawnpore Karmachari Union founded in 1954. In addition, there were *seven* other unions in

Table 2.3 ***Size and Rate of Cancellation of Trade Unions in Kanpur in 1976***

Size of Unions (Excluding Federations)		*Number of Unions*	
		In Existence in 1976	*Created since 1966 but not in Existence in 1976*
0–50	Total	67	15
	Aff.	14	3
51–150	Total	96	30
	Aff.	37	15
151–300	Total	73	23
	Aff.	22	12
301–600	Total	44	7
	Aff.	22	2
601–1,200	Total	45	7
	Aff.	27	4
1,201–2,400	Total	21	2
	Aff.	15	2
2,401–4,800	Total	12	2
	Aff.	7	1
4,801–10,000	Total	4	0
	Aff.	4	0
10,001 +	Aff.	1	0
Total		363	86

Note: Aff. indicates unions affiliated to federations, and are included in the totals in the line above.

Source: Registrar of Trade Unions, Kanpur. Collected by the author from registration applications.

Table 2.4 ***Trade Unions in Kanpur by Size and Affiliation to All-India Federations***

All-India Trade Union Federations	*Size of Local Unions*			
	Up to 300	*301–3,000*	*3,001 and Above*	*Total*
Indian National Trade Union Congress	15	15	4	34
Hind Mazdoor Panchayat	8	5	1	14
Bharatiya Mazdoor Sangh	39	20	1	60
Hind Mazdoor Sabha	5	6	1	12
United Trade Union Congress	9	10	1	20
Uttar Pradesh IMF	3	2	–	5
Confederation of Indian Trade Unions (Communist)	5	3	–	8
National Labour Organization	–	2	–	2
Rashtriya Mazdoor Federation	1	1	–	2
All-India Trade Union Congress	1	–	1	2

Note: There were also the following federations consisting of clerks:

Uttar Pradesh Audyogic and G.S. Associations	5 unions
Uttar Pradesh Bank Clerks' Federation	3 unions
All-India Bank Employees' Federation	1 union
Uttar Pradesh Bank Employees' Federation	1 union

Source: Records of the Registrar of Trade Unions, Kanpur.

the company. The full time vice-president of the union was formerly a senior *mistri* in the carding and spinning section. He became a full time trade union executive member because he was afraid he might lose his job because of his previous record of violence. This union had sometimes been affiliated to the INTUC and sometimes not, but since it wishes to represent all the workers in the company, the officials do not exercise any political pressure as a union on its members to vote for any political party, although they claim to give 25 per cent of union money to whichever federation they are affiliated to. Being a large union they are mostly concerned with problems of wages and payments. However, at the time I was there, the main problems were a fight over the wages for a new wool-scouring machine operator and over bonus, which was a wages issue. Under the Uttar Pradesh industrial law, a firm is not legally responsible for paying bonus if it showed a loss the previous year. This is often a matter of accountancy.

Of the seven executives of a union, three may be non-members of the firm. Naturally, such executives cannot be sacked by the firm, thereby losing their executive position on the union, as they are not employees. In many cases they are either persons who make their living as labour consultants (who may by law not be lawyers) or are organizers in one of the federation trade unions. These persons make their livelihood by negotiation. The most common method of remuneration is to receive 10 per cent of the money received by workers in back pay in a dispute, plus a daily allowance and travelling expenses for each day spent in negotiation. I have met a number of these negotiators. The most normal pattern for those who are not employed by a federation is that they were former trade unionists who became discharged for being activists or for using violence on the employers—that is, for being good organizers. They often formed unions with their former workmates and negotiated with their former employers. For the most part, I believe, they had the interests of their union at heart. Two of the most well known organizers received Rs 2,500 and Rs 70,000 respectively for wrongful dismissal from their firms and set up as labour advisers. The first represented a union in a fertilizer factory for some years and then switched over to representing the employers in labour disputes. The other set up as a labour adviser and then established his own factory which was well known for exploitation and sharp practices. For the most part, however, I was impressed with the ability and sincerity of these organizers. There is only one major crime for these labour advisers—accepting money clandestinely from the employers while representing a union to call off a successful agreement at a lower rate of increase.

In the case of large unions (which perhaps may organize as much as 45 per cent of the workforce in a firm), the most important issues which they concern themselves with are increase of the total wage packet, and negotiations about bonus in firms which are showing little profit. If they are successful in keeping their supporters together, they may make a successful income as a much higher proportion of the *increase* in wages goes to union funds than from the original wage. But those disputes which deal with a collective increase of wages can usually only be dealt with by those unions connected with a federation as it is the political power of the federation which enables such a dispute to be successful. This is why the proportion of unions attached to federations in Table 2.3 increases proportionately as the size goes up. This also explains why only 158 unions are listed in Table 2.4 whereas there were 449 unions registered at this time with the Registrar of Trade Unions. Almost two-thirds of the unions were not connected with any form of federation, although practically every one of these unions had as an adviser on their executive some person not employed by the factory.

The majority of unions do not receive even the minimum Rs 3 per annum from their members, and official accounts are certainly rigged for the most part as membership figures are not checked against financial contribution by any government agency. It appears that about 10 per cent of the workers belong to more than one union. (This figure has been disputed and is difficult to verify.) But in a large textile or ordnance firm the unions become very bureaucratized and it is the small issues which are often of more importance. If an individual has a dispute of some kind, he will look for a union which will support him. The following smaller disputes were being dealt with by an organizer attached to a Jan Sangh federation in one ordinary week.

1. A termination dispute. A fitter in the New Victoria Mills claimed that he should have been promoted and won Rs 3,200.
2. An employee in Hindustan Automobiles had his services terminated and received Rs 21,400.
3. Two dying core operators in Muir Mills asked for a rate revision case for their work and gained an increase of Rs 10 per month.

These are the types of disputes which individuals use unions for, and are usually the meat for small unions to develop on if a large union refuses to negotiate such small individual disputes.

When a major bonus or salary revision case is under way, however, the existence of numerous smaller unions, which are naturally much more militant as they have nothing to lose, make it difficult for both employers and larger unions to deal with each other, as the small unions have a status equal to the larger unions as independent parties. It is not uncommon for unions to come into violent conflict with each other, and officials of unions are often beaten up by other union workers (Ramaswamy 1977).

However, within the firm one gains substantial status out of being on a union executive. It is impossible in a firm such as the Woollen Mills to move up in working status after one has been classified in one's category. One cannot alter one's position in the whole system. One cannot change, say, from a washer in the finishing section to a dobber in the weaving section; so membership of a union executive is the most important means of obtaining further upward mobility in status within the company. In the Cawnpore Woollen Mills there were seven unions, meaning that there were at least 35 union positions available instead of five. Since every union was legally equal to negotiate with the management, provided that it could recruit the requisite minimum 10 members, the smaller unions would take up the problems of individual members with the management and thus acquire prestige, whereas the larger bureaucratically organized unions dealt with the more general questions of a general increase in wages and bonus. It seems that unions below a certain size pass through a recognizable cycle based upon changes in the nature of support they receive from differently aged members. This cycle has already been analyzed for Japan by Wakao and Funabashi (1960). But in certain types of Indian firms, I believe, his ideas can also be shown to apply.

In short, the excessively high number of classifications of job categories left over from a previous industrial era which no longer represent real skill differentials, plus the relatively high dearness allowance in proportion to wage differences, prevent wages or job classification becoming a basis of status differences. In the mill, the 300+ job classifications were too numerous to form the basis of a status system. Moreover, since dearness allowance was greater than wages, differentiation in income was slight. A 10 per cent (Rs 5) difference in an income of Rs 50 becomes unimportant in proportion to an income of Rs 500. So wages decreasingly become a means of status differentiation *within* the factory. In addition, since each section of the mill was autonomous as far as promotion was concerned, once one had attained a permanent position, there was no further way one could receive promotion within the firm itself. Yet one only had to be within the Cawnpore mill for a comparatively short while to realize that the Indian worker was extremely status conscious, from the graduate engineers at the top who always wore a tie even when undertaking dirty work, to the difference in intonation in command whether one was asking a coolie or a fellow weaver to help in some task. Moreover, it was persons of the same status who usually ate together in the off breaks. If one were a member of a union, even a small one of a few tens of members, everyone normally addressed members as Mr Secretary, Mr Treasurer, and so on. The fact that an executive member of a union could not be easily discharged under the Uttar Pradesh labour law is by no means a privilege which could be ignored. The argument that I am putting forward here therefore is that the more unions there are, the more status mobility is possible in the case of status differentiation. Moreover, the larger unions are more bureaucratized in the sense that they show less interest in taking up the cases of *individual* workers. Caste as a sign of status is of little importance inside the mill, but the attitudes associated with caste are to some extent transferred to the work situation, to membership of trade union executives, to one's rank inside the section of the factory where one works, and to the length of time one has remained inside the mill, which is closely associated with age because of the lack of upward social mobility and permanence of employment.

I shall, on another occasion, describe the situation in one of the newer engineering skilled firms established since the war. I will only say here that the different working conditions, the much easier and almost automatic way skilled workers in a factory can gain increased status with increased skill on the machines, and the emphasis on ability make the situation in the mill and in an industrial factory entirely different. For example, on the application form for appointment in engineering firms, there is no space for caste or religion of the applicant but, in the mills, caste and religion are placed immediately after the name and address. The different situation also affects trade union membership. For instance, the supporters of trade unions usually either consist almost entirely of unskilled workers or almost entirely of skilled workers, but it is unusual to find an engineering firm in which all the permanent workers, whether skilled or unskilled, give equal support to the union in their firm if one is formed.

I only wish to emphasize that in the work situation, those features of the wider Indian culture and society which are emphasized vary. How they vary form the basic characteristics of the new branch of social anthropology—industrial anthropology.

Practically every book that Srinivas has written has as its background caste in some form. Residence, the organization of work, the ownership of property—nearly every aspect of Indian life in Srinivas' books deals with caste. What Srinivas has done in his various articles is to show that caste is so flexible and so important in Indian society that any attempt to reduce it to any other sociological category (such as class or hierarchy) over-simplifies its nature. Nearly all Srinivas' books and articles refer to the rural scene, and when Srinivas deals with caste as a basis of provincial organization, as in his study of Karnataka politics, he has no difficulty in showing that caste associations still retain marked caste features, even when these associations no longer have characteristics such as endogamy but become merely a means of organizing power blocks together (see Srinivas 1957).

In this article, I submit, work and the way it is organized within different sorts of factories are primary determinants of status within the factory. Since the excessively numerous job classifications in the Lal Imli factory no longer represent real differences between the different classes of workers, any attempt to try and differentiate between workers makes the trade union system (which by law has members only from one factory) become a forum in which different individuals try and gain increased respect from their fellow workers.

The dominant caste in a village has behind it the full force of the Hindu religious system, the ownership of land, and other signs of authority. But in a factory, where the majority of the workers do not think either Hinduism or the ownership of capital is especially relevant to the exercise of authority inside the factory, the use of the traditional system of control is likely to be unsuccessful in the long run unless it also clearly becomes the most effective way of organizing the productive system. The caste system trains people to understand the basis of inequality by supporting different statuses, but because of the inflexibility of the system in Lal Imli with its different sections, its countless classifications and its lack of mobility with little actual wage differentiation due to dearness allowance, the trade unions have become one means by which one can gain status and respect inside the factory. This membership of a trade union executive is uncontrolled by the formal structural system of the Lal Imli management, which has become encapsulated within its hundred years of history and its now useless job classification system.

REFERENCES

D'Souza, Victor S. 1975. 'Scheduled Castes and Urbanization in the Panjab: An Explanation', *Sociological Bulletin*, 24, 1: 1–12.

Ghosh, Saila K. and Sengupta, Amil K. 1978. 'Cancellation of Trade Unions and its Implications: The Case of West Bengal', *Decision*, 16, 4.

Holmstrom, Mark. 1976. *South Indian Factory Workers*. Cambridge: Cambridge University Press.

——. 1981. 'Factory Workers and "Unorganised" Labour in India', paper presented at the Seventh European Conference on Modern South Asian Studies, London.

Pandey, Gyanendra. 1990. *The Construction of Communalism in Colonial North India*. Delhi: Oxford University Press.

Pandey, S. M. 1970. *As Labour Organises*. New Delhi: Sri Ram Centre.

Ramaswamy, E. A. 1977. *The Worker and His Union*. Delhi: Allied Publishers.

RAMASWAMY, UMA. 1983. *Work, Union and Community: Industrial Man in South India*. Delhi: Oxford University Press.

REX, JOHN and MOORE, ROBERT. 1967. *Race, Community and Conflict: A Study of Sparkbrook*. Oxford: Oxford University Press.

SINGH, H. H. 1972. *Kanpur: A Study in Urban Geography*. Varanasi: Indrasini Devi.

SRINIVAS, M. N. 1957. 'Caste in Modern India', *Journal of Asian Studies*, 16, 4.

WAKAO FUJITA and FUNABASHI NAOMI. 1960. *Nihongata rõdõ kumiai to nenkõ seido*. Tokyo: Tõyõ Keizai Shinpo-sha.

3

World-Views and Life Worlds: Case Studies of Industrial Entrepreneurs in Faridabad

M. N. PANINI

This study[1] is concerned with the world-views of industrial entrepreneurs[2] in Faridabad and relates them to their respective life worlds,

[1] This paper is based on the research work reported in my Ph.D. thesis entitled 'A Sociological Study of Entrepreneurs in an Urban Setting', submitted to the University of Delhi in 1979. The research was conducted under the supervision of A. M. Shah, to whom I owe more than I can acknowledge. My fieldwork in Faridabad was conducted during the year 1973–74. It is important to keep this in mind in view of the fact that industrial policies have been liberalized substantially since that time. My interest in this topic was kindled by Singer's pioneering work on the industrial leaders of Madras. I have sought to raise issues similar to the ones that Singer raised, but I do so within the framework of Berger's sociology of knowledge perspective. This perspective seems to me to be particularly useful as it articulates world-views with the social context in which they emerge and operate.

[2] From the Bergerian perspective the English term 'entrepreneur', which is used as a self-definition by the owners and heads of industrial establishments in Faridabad, is

viz., the social milieux or contexts in which they grow up, live and work. The term world-view refers to cognition of the mundane world and the sacred realm; it incorporates the way in which the mundane and the sacred worlds are comprehended and the way gods, people, events and things are accounted for, interpreted and evaluated.[3] Defined thus, this paper encompasses the Weberian theme of the role of values and religion in development. It can be inferred from the problems posed here that my main concern is not *per se* with the analysis of values and beliefs as expressed in the scriptures, but with the way people interpret and use them in relating themselves to their life worlds.

FARIDABAD INDUSTRIAL COMPLEX

In an earlier paper on industrial entrepreneurs in Faridabad, I had provided details of the Faridabad township and the economic relationships governing the light engineering goods industry which had been selected for intensive study (Panini 1978). Hence, I shall recapitulate here only some of the more salient details. Faridabad has, in the last three decades, gained prominence as a leading industrial centre in the country. Its proximity to Delhi, which is about 20 km to its north, and its location on the important rail and road routes of India, have contributed to the rapid expansion of this town.

Faridabad town belongs to the former Gurgaon district[4] of Haryana. The town was selected as the site to rehabilitate refugees from the North West Frontier Province Agency of Pakistan (NWFP for short) in the wake of Hindu-Muslim riots following the partition of India in 1947. A new township was developed on the outskirts of the old town to settle the refugees, and the government sought to promote industrialization to provide employment to them. Faridabad has come a long way since those days; known only for its *mehendi* (henna) plants before Partition, it has now become a flourishing industrial centre. As

---

most appropriate for my study. This usage does not commit me to accept entrepreneurs as innovators in the Schumpeterian (1934) tradition, although some executive entrepreneurs tended to be innovative in seeking solutions to their problems.

[3] In short, I am using the term world-view to refer to what Berger calls cognitive style. See Berger (1973).

[4] The Haryana government has recently created a new Faridabad district with the town as the headquarters.

one leaves Delhi for Agra in the south, ribbons of industrial units straddle the road and rail tracks till one travels beyond Faridabad.

Faridabad's industries have been attracting a huge stream of migrants from various parts of the country. The town is the home of entrepreneurs from almost every state in India, especially Punjab (including Punjabi refugees from Pakistan). The NWFP refugees have been mostly small shopkeepers, contractors, brick-makers and industrial workers. They form a significant voting bloc in the town politics and have dominated the local Municipality. They are, however, no match for the Punjabi entrepreneurial elite who, some years ago, exerted pressure on the state government to get the Municipality suspended in order to improve the civic amenities in the town. The Administrator of the township demolished numerous *khokhas* (temporary wooden structures used for petty trade and as tea shops) to widen some important roads in the town, a task the elected Municipality could not undertake owing to pressure from the NWFP refugees who owned most of the *khokhas*.

Although the Punjabi industrial elite of Faridabad is highly influential at the state and central government level, the levers of power at the district and state levels are controlled by the landed castes of the region, viz., the Jats, Gujjars and Meo Muslims.[5] These landed castes have been strong enough to pressurize the State government into diverting electricity to agriculture at subsidized rates even during periods of acute power shortage, resulting in the closure of many industrial units. This rural elite, however, is not interested in extending its sphere of influence to the industrial world of Faridabad.

The Faridabad industrial complex can be described in terms of vertical and horizontal ties that connect the various industrial units. Vertical ties consist of job work and labour job relationships. Job works are contracts for particular industrial jobs handed out by large industrial corporations to small industrial units. The raw materials required for the job are to be procured by the vendor units. The vendees provide detailed drawings and technical help. Being a vendor to a reputed industrial corporation helps in procuring import licenses, raw material quotas and concessional loans from commercial banks and other financial institutions. Job works provide an assured sales outlet at rates fixed in advance, eliminating thereby a major source of uncertainty and anxiety. A negative feature of the relationship is the

[5] See *The Times of India*, 12 December, 1984, 'Faridabad Voter is Unpredictable'.

pathetic dependence of the vendor on the vendee firm. The vendor, having invested a large amount in the development of tools required specifically for the given job, finds himself strongly bound to the vendee. The vendee can control the vendor by raising the percentage of items rejected for not meeting quality specifications and by delaying payment for the items delivered. As there is severe competition for vendorship, the vendee firm can virtually dictate terms. Hence the vendors who find the relationship too oppressive seek to set up a tool room to concentrate on repair and maintenance work of various industrial units. As this involves much skill and ingenuity, the entrepreneur can quote his own rate.

The vendee units gain in a variety of ways from job works. They can expand their scale of production utilizing vendors' investments. This allows them to skirt 'labour problems' like strikes and industrial disputes, which are inevitable when a large workforce is employed. They also save on Provident Fund payments, health insurance and other statutory benefits for employees. The vendees can concentrate on the more sophisticated and difficult jobs, leaving the peripheral jobs to the vendors. They can also pass the cost of high quality to the vendors. This institution helps the vendees to project themselves as the promoters of small industries development in the country, which is in line with the government's industrialization policy. Job works, therefore, enhance the patronage of big firms and add to their prestige in society.

Labour jobs refer to informal relationships in which the vendor is supplied with raw materials as well because he lacks capital. The labour job vendors are former skilled workers who set up a workshop with an old lathe or some other such machine and install it in their own backyards, as it were. Labour job vendees are quite often vendors of job works. Labour jobs are known to be less profitable than job works and are indeed labour intensive. Although the vendors of labour jobs look upon themselves as independent entrepreneurs, they can be described as disguised workers.[6] The relationship between the vendee and the vendor of labour jobs is reminiscent of the patron-client relationship, with its emphasis on personal trust. Hence, labour jobs are not easily available.

The vertical ties described in the foregoing are criss-crossed by horizontal ties of equality. Job work relationships expressing greater

[6] Harriss' description of the industrial structure of Coimbatore (1982: 945–54, 993–2002) runs close to my description (1978) of the Faridabad industrial structure.

equality resemble labour jobs. Vendee firms are particular that their sophisticated jobs, using special raw materials, be entrusted to known vendors who have the technical competence and the machines required for the job. The vendees in such cases may even provide the required raw material if it is not easily available. Horizontal ties are relationships of mutual dependence. Such relationships exist among friends, complementarity rather than competition being the hallmark of friendship.

Vertical ties stratify entrepreneurs in the engineering goods sector into four categories. At the top are the industrial leaders who control firms and corporations, some of which have acquired a national, if not an international, reputation. They are the vendees who hand out job works. They possess management experience either in working for the government or for a private firm. Some of the industrial leaders are dynastic entrepreneurs who took over the reins of entrepreneurship from their fathers; others shifted to entrepreneurship from industrial management and government administration. All of them are graduates, possessing degrees in engineering and management. The exception to the rule was the son of a leading industrialist who joined his father's firm in a managerial capacity soon after school. Sharing the experience of management is another category of industrial entrepreneurs, who are termed executive entrepreneurs (executives, henceforth). This category consists mostly of vendors of leading industrial corporations. Most of them head small-scale industries. There are some executives who have been highly successful and have expanded their firms beyond the investment ceiling for the small-scale sector.[7]

There is another category of entrepreneurs who depend on job works. They are termed supervisory entrepreneurs (supervisors, henceforth) because their previous work experience had been as supervisors in industrial firms.[8] This is a mixed category. There are some with a

[7] The government's definition of a small industrial unit has been changing over the years, with the ceiling of investment in plant and machinery being continually pushed up. At the time of research the investment ceiling was Rs 7.5 lakh, and Rs 10 lakh for ancillary units. Now the ceiling is Rs 35 lakh and Rs 45 lakh respectively.

[8] There are two cases of supervisors who did not possess prior industrial work experience. They invested in industry for reasons of prestige. One was a contractor. He suffered huge losses in the initial stages of his industrial entrepreneurship. Subsequently he made it good thanks to a neighbouring supervisory entrepreneur. The latter helped him tide over his initial crisis. The other was a taxi-operator. He entered into a partnership with a supervisory entrepreneur before buying over the entire workshop.

few years of formal schooling who start as industrial workers and get promoted to supervisory positions. There are also younger persons who begin their careers as supervisors after completing a diploma course in engineering. As competition for job works is severe, they depend mostly on labour jobs while a few manage to become vendors of job works. Below the supervisors are the craftsmen entrepreneurs (craftsmen, henceforth) who depend on labour jobs. With only a few years of formal schooling, if at all, they receive early training in industrial work as helpers of skilled industrial workers, whom they reverentially call *ustad* (master) in Urdu.

The industrial leaders, executives, supervisors and craftsmen form a rank order of status. The industrial leaders and executives hail from the upper or middle strata of society. They belong to families of industrial entrepreneurs, businessmen, government servants or professionals. Some of the older executives are sons of landowners. Many of them gave up attractive careers as industrial managers, both within the country and abroad, for the independence and creativity offered in industrial entrepreneurship. The supervisors belong to the middle or lower middle strata, their fathers being in middle or lower level urban occupations were policemen, clerks, school teachers, small business men or artisans. The craftsmen's parents were artisans, industrial workers, small cultivators, or agricultural labourers.

As will be demonstrated subsequently, particularistic criteria (such as caste and religion) are significant for the world of industrial entrepreneurship. Hence, it is important to turn to a brief description of the caste and religious backgrounds of industrial entrepreneurs. The caste composition of the fifty entrepreneurs selected for intensive study[9] reveals interesting continuities as well as breaks with hereditary occupations. Artisan castes (such as Dhiman, Panchal, Ramgariah, Saini and Sonar) are represented in the categories of craftsmen and supervisors, indicating continuity with their traditional occupations and lack of access to modern education. Those belonging to the trading castes (such as Agarwals, Khatris and Aroras) dominate the world of industrial entrepreneurship, showing thereby that they have found the shift from trade to industry easy. The trading castes are roughly equally

[9] The selection of these entrepreneurs was based on the networks prevailing among them. That is, I asked each entrepreneur I met to introduce me to his other entrepreneur friends. This method proved useful in outlining the industrial structure of Faridabad, which would not have been possible using the survey method. For details see Panini (1979).

distributed among all the entrepreneurial categories, indicating that some have availed of the facilities of modern education to get a better start in life as compared to the others. The Brahmin, Rajput, Jat, Kayasth and Amil (Sindhi) castes are represented mostly in the top two categories, indicating the mobility of upper castes through modern education. These upper caste entrepreneurs in many cases have been first generation entrants into the field of industry and commerce. They also hail from distant states in India, which is also an adequate commentary on their geographic mobility.

Apart from the Hindus, Faridabad has attracted members of the Sikh and Jain religious communities. The township has not encouraged industrial entrepreneurship among the Muslims, as indicated by the *Who's Who of Faridabad Industries* (Soni 1973). My research strategy succeeded in drawing a lone Muslim entrepreneur into my sample.

In an earlier paper (1978), I have argued that the hierarchy of industrial entrepreneurs in Faridabad represents a system of cumulative inequalities. The craftsman entrepreneurs cannot develop beyond the investment threshold[10] of around Rs 50,000.[11] Lack of education and formal training prove to be severe handicaps in coping with the complexities of management and formal existence as independent firms. The supervisors whose threshold is the ceiling limit for small-scale industry, however, are capable of learning new ways of adapting themselves to their milieu. This is particularly true of the diploma-holder supervisors. In the case of executives and industrial leaders, however, such thresholds to growth[12] cannot be observed.

In the paper (1978) mentioned above I related the styles of the four different categories of entrepreneurs with their development patterns. Entrepreneurial styles refer to the typical manner in which different categories of entrepreneurs relate to their significant others, organize work and deal with the accounting and technical problems involved in managing an industrial unit. In this paper, I have studied the correspondence between world-views, entrepreneurial styles and the development pattern of the different categories of entrepreneurs in

[10] The term threshold simply refers to the hurdles in entrepreneurial careers. It is not used here in the Van Gennepian sense (1960: 20).

[11] The figures refer to the 1973–74 period. See Panini (1978: 102).

[12] Thresholds restrict. The absence of thresholds does not, however, guarantee success. The lack of motivation, ill health and a host of other factors may block one's path to success. As will be shown subsequently, those who do not possess thresholds to growth are capable of recovering even from a near failure situation.

order to obtain a comprehensive understanding of linkages between industrial entrepreneurship and social change.

THE CRAFTSMEN ENTREPRENEURS

The craftsmen begin their entrepreneurial ventures in a small way owing to the paucity of capital. As industrial workers, they cannot save large amounts for investment. They rely mainly on the Provident Fund and gratuity they get when they retire or resign from their jobs. The fortunate few who can raise money from relatives and friends may try to set up a workshop, even while working full time in a factory. Their initial investment is usually on a low priced lathe or an old machine which they repair and use. Their workshops are often located in their own houses as they cannot, in the initial stages, rent or buy an industrial shed. Some shrewd craftsmen possess no machines but hire them from friends.

The craftsmen face heavy initial odds because of keen competition among hundreds of skilled workers aspiring to become independent as industrial entrepreneurs. To attract labour jobs, they have to quote low rates, ensure a high standard of work and be prompt in meeting deadlines for delivering the items. It is, therefore, common to find craftsmen working for long hours on their machines. Because they quote low rates, they attempt to reduce costs in every possible way. Much care is paid to recovering every bit of scrap in the premises and to the use of machines. To reduce machine use, they seek to maximize the use of hand tools, although this may often involve long and laborious work. Some develop devices which could be attached to their machines and modify their machines to increase productivity. To expand their operations, they seek to maximize their savings by living extremely frugally.

Despite such attributes, many craftsmen fall by the wayside, as it were. A select few, however, weather the initial crises and even grow to establish a local reputation for skilled work. Their survival depends, among other things, on their skill and ingenuity, and capacity for improvization, and on their vendees who protect and nurture them. Here, Bhatia's example is pertinent. He once heard through his friends that an industrial corporation was auctioning a large amount of steel scrap. He raised some money from his relatives to bid in the auction successfully, and then he promptly set out to convert the scrap into bolts and nuts. He sold them to dealers at competitive prices and managed to make a tidy profit. But the protection that vendees can provide and their ingenuity prove inadequate when their entrepreneurial growth reaches an investment threshold of Rs 50,000. As they near the threshold, the complexities of workshop organization and management multiply. As they buy more machines, they can no longer rely on labour jobs alone to fully utilize the machines. They find it difficult to take up job works because of the shortage of working capital and because they cannot afford to wait for over three months for the payment of their bills. The craftsmen realize, as they expand their unit, that there is more to entrepreneurship than merely working with machines. They can no longer work unobtrusively in their own houses. They have to shift to a proper industrial shed and fulfil various legal formalities. They have to obtain a licence from the Municipality to operate and obtain proper electricity connections from the Haryana State Electricity Board. They also have to cope with factory inspectors, sales tax and income tax officers and such others who may come on periodic inspection. Having little knowledge of official procedures and record keeping, the craftsmen are easy prey to unscrupulous government personnel. Hence, they prefer to avoid officials whom they regard as 'thieves'.

The craftsmen face problems not only in coping with the milieu external to their workshops but also in organizing and managing their internal affairs. As the workshop expands in size, it is no longer possible to depend on simple calculations to estimate profits. Profit and loss are no longer simple notions based on the difference between receipts and payments. They need to use more sophisticated notions of profit and loss and, in order to calculate them, they have to keep proper accounts and be able to do some abstract calculations. Further, to avoid harassment from sales tax and income tax officials, the ability to doctor accounts is an advantage. They have a strong antipathy towards 'paper work' and are unable to successfully cope with the changing *gestalt* they face.

The delineation of the development pattern of the craftsmen in the foregoing illustrates the struggle that the craftsmen have to put up in the competitive world of entrepreneurship. Despite heavy odds, the craftsmen do not give up easily. To understand what sustains them in their struggle, it is necessary to grasp their hopes, ambitions and values, and the way they look at themselves and at others—that is, their world-view. A brief description of the world-view of the craftsmen follows.

The craftsmen share the general ethos of the Faridabad industrial world, which values independence rather than doing 'service', or work, for others. Many craftsmen and industrial workers in Faridabad nurture the ambition of becoming big industrialists, producing one's 'own item', that is, manufacturing an industrial component or machine selling under an exclusive brand name. Craftsmen admiringly refer to some rags-to-riches examples in the township and hope to emulate them. These stories, when verified, turn out to be myths; but they sustain the craftsmen nevertheless.

While valuing independence, the craftsmen do not look at all occupations providing independent livelihood as equally attractive. While being an industrial entrepreneur is considered prestigious, the work of a *kabadiwala* (dealer in old newspapers and junk) or *panwala* (seller of *pan*, betel leaves) is considered to be of low prestige even if it fetches more money. In fact, it is this prestige dimension that drives many people who have made money elsewhere to invest in industry. There are two entrepreneurs in this study who entered industrial entrepreneurship solely for reasons of prestige.

The craftsmen are proud of their industrial skill. They say that since they possess 'art' in their hands, they are confident of earning a livelihood in any eventuality. They often recall incidents when they solved mechanical problems which baffled even trained engineers. They are particularly fond of saying that they can do mechanical jobs which are not done anywhere else in the country. Some of the senior craftsmen talk about their reputation as skilled workers for skill being so big that even large and reputed companies may often entrust difficult jobs to them.

Pride in mechanical skills is accompanied by a disdain for what they call 'paper work', involving writing and calculation. They even imply that 'paper work' involves cheating and cleverness, in contrast to manual work which is honest and forthright. Dignity of labour indeed works with a vengeance among craftsmen.

In spite of their skill and ability to work hard, the craftsmen know that they face obstacles in becoming big entrepreneurs. Having dealt with corrupt government officials who book them for violating one or another law or regulation unless given a bribe, the craftsmen have a ready explanation for the struggle they have to put up in order to succeed. Easy success is regarded as synonymous with corruption. They point to some leading industrialists in the town and attribute their success to their ability to 'eat and drink' with government officials and others. To 'eat and drink' implies entertaining on a lavish scale, with alcoholic drinks and sumptuous dinners. Some of them even suggest that successful entrepreneurs 'prostitute' their wives. 'Look at the way their women dress and behave so freely with other men. Aren't they loose?' they ask. For craftsmen, to take one's wife out for dinner and allow her to mix freely with other men is to encourage her to be 'loose'.[13] This they regard is the way to develop 'approach', by which they mean cultivating ties with influential people. The craftsmen project themselves as moral beings who, because they refrain from 'eating and drinking', face many hurdles in the path of success. They say that they have to indulge in the evil of giving bribes to government officials to avoid harassment. They cannot understand why government officials cannot leave them free to pursue their occupation. After all, they are earning an honest livelihood!

The moralistic stance that the craftsmen adopt in evaluating success reflects one facet of the social world they construct. At home the craftsmen are in a world of personal relationships consisting of kin and affines, vendees of labour jobs, and friends. They can best relate to others in terms of the notion of 'honour'.[14] Thus, *vis-à-vis* their kin, the craftsmen adopt the norms of patriarchy. Among the 12 craftsmen, eight are heads of their own households and the remaining are bachelors living with their widowed mothers, elder brothers and elder brothers' wives and children. As heads of households, the craftsmen regard themselves essentially as providers of the family, and expect their wives and children to adjust themselves to the exigencies that may arise in the workshop. The wife is called upon to take on the extra burden of the husband's chores (like shopping) because the husband is kept busy in the workshop. The craftsmen realize their obligations to their wives and children but plead helplessness in fulfilling commitments like taking the family out to the cinema. To compensate for their neglect of family obligations, a few craftsmen have installed a television at home. While the craftsmen who expect deferential behaviour from their wives and children they are, in turn, respectful to their elders. The elder brother, who is respected,

[13] The few parties and dinners given by executive entrepreneurs to which I was invited were a big contrast to the free-sex revelry that craftsmen imagine them to be. Women and men tended to remain segregated most of the time. They were dull and formal affairs, with men talking shop most of the time.

[14] See Berger (1973: 78–83) for the distinction between 'honour' and 'dignity'.

is seen as a protective and disciplining authority. The craftsmen do not find it easy to violate the advice of the elder brother in both family matters and workshop affairs. Although traditional notions of kinship and marriage inform craftsmen's lives, a change can also be noticed. Obligations emanating from extended kinship and marriage ties no longer remain significant. Work involvement prevents them from attending, say, a marriage of a distant relative in a nearby town or visiting some distant kin member residing in a different part of the town during a festival. Occasionally, if such a relative also happens to be an industrial entrepreneur or an industrial worker, an intimacy develops because of common concerns. In general, however, there is a tendency towards narrowing the circle of kinship and marriage.

The work-centred world of craftsmen is filled mostly by intimate and informal relationships with patrons, *ustads* and friends. Their patrons are vendees of labour jobs. The mutual trust that marks the relationship is built up over several years of close association on the shop-floor of a factory or a workshop. The patrons may have been engineers or officers under whom the craftsmen were placed in their previous employment. Invariably, the craftsman would have helped his senior colleague in branching out as an independent entrepreneur by assisting him in setting up the machines and procuring reliable workers for him. The craftsmen may be occasionally called upon to substitute for an absent worker when there is a deadline to be met. The patron reciprocates by lending some old machine or machine tool when required, by using his influence among his friends to procure labour jobs and by occasionally mediating with government officials on behalf of his client.

Mutual trust obtains in the intimate relationship between the craftsman and his *ustad*. The *ustad* is acknowledged by the craftsman as the master under whose tutelage he learnt his industrial skills. The *ustad* helps by lending his protege money required for setting up the workshop. He may also use his influence among senior workers in various small-scale industries to procure labour jobs for his protege. An example is the instance of a senior and respected worker in a small workshop who used the trust he had built up with his employer to divert some labour jobs to his protege or *chela* with whom he had entered into a partnership. This relationship is shot through with notions of loyalty, trust and reverence towards one's *ustad*.

The craftsmen are most free and relaxed in the company of their friends. Friends are former work-mates, some of whom may also have taken to entrepreneurship. Friends drop in at the workshop off and on. They bring news of old machines on sale, market trends, profitable lines of investment, and gossip. Friends having two different machines may share labour jobs which call for the use of both. A craftsman may recommend his friend for a labour job requiring a special machine. One craftsman kept accepting labour jobs even after he had sold his lathe because he was confident of hiring one from his friends.

Relationships with friends come nearest to being relationships of equality. Equality is expressed in terms of the liberty they take in trading abuses and sharing jokes and gossip. Nevertheless, friendship gets strained when the friend becomes a potential competitor. The spontaneity and freedom that marked the relationship earlier is replaced by caution and suspicion. It is complementarity rather than competition that binds friendships.

The craftsmen operate in a social world of multiple castes. They can establish relationships across caste and religion provided the idiom of 'honour' is carried over. That is, so long as inter-caste relationships are expressed in terms of patronage and clientship, they find it easy to deal with them. Caste and religion are, however, not rejected. Far from it; if a friend or patron happens to belong to the same caste, it contributes to a closer and more intimate relationship. For example, a Ramgariah Sikh craftsman would approach other Ramgariah industrial entrepreneurs for labour jobs with a degree of self assurance about the outcome absent while approaching others.

The craftsmen extend the idiom with which they relate themselves to the social world to their relationship with god. God represents a more powerful though invisible patron. One worships and makes offerings to god as a gesture of reverence and gratitude. God is viewed as intervening in human affairs to reward and punish. When I asked Jagir Singh, a Sikh craftsman, why he ritually throws a few coins in his furnace everyday before he begins his foundry work, he replied that it was his offering to god for filling his stomach everyday. All the craftsmen, indeed all the entrepreneurs from the most humble to the very sophisticated, displayed pictures of Vishwakarma, the engineer-architect in the Hindu pantheon, in their workshops and offices. Apart from Vishwakarma, they display pictures of other gods or *gurus* depending on their religion and family tradition. The Sikhs displayed pictures of Guru Nanak, Guru Govind Singh and other *gurus* alongside Vishwakarma's portrait. The Hindus displayed pictures of Hindu gods and godesses, the most popular being Hanuman and Lakshmi. All of them begin the day by cleaning the table they use and the shelf or altar they have set up in front of the god's picture. Then they offer worship by placing flowers on the pictures and lighting incense. Some do the same for the table and the account books they keep. Even the seat from which they transact work in the workshop (that is, the *gaddi*) is worshipped. *Gaddi* is regarded as the seat in which the goddess Lakshmi is supposed to reside. Bhatia, a craftsman, was once describing how the entrepreneur next door was a double dealing and non-cooperative person, but he quickly checked himself saying that he should not be talking ill of others while sitting on the *gaddi*.

The craftsmen are not particularly ritualistic, but they regularly visit temples and *gurdwaras* and undertake pilgrimages. Some of the Hindu craftsmen visit *gurdwaras* while a Sikh craftsman reported having been on a pilgrimage to the Hindu temple of Vaishno Devi in Jammu. Two Hindu craftsmen are on the temple committee in their locality, and one is also a member of a voluntary religious body which conducts Ramlila celebrations in Faridabad.

The craftsmen interpret religion, be it Hinduism or Sikhism, as teaching one to be good and honest. One should remember god's name and should not harm or cheat others. God is seen as punishing those who do something wrong. If one's business is not prospering, one has to win god's favour through worship and good conduct. They use the term *dharam* (colloquial for *dharma*) multivocally. It may mean religion in one context and morality in another. They do not use the notion of *karma* as often, although in their daily conversation they often make references to rebirth. Thus a craftsman may say, when discussing his disappointment with his son taking to bad ways, that maybe in his past life his son had been his enemy. They regard the existence of god as a self-evident truth. When I asked them questions about whether they believe in god, they took it to mean that I doubt the existence of god. They can understand my believing in another god because I was different from them but how could any man be godless! To be godless amounts to not being human. Some have come across communists in trade unions who do not believe in god. According to them that is the one thing that is wrong with the communists.

The craftsmen are not too involved in giving elaborate interpretations of religion. For instance, a Ramgariah Sikh craftsman directed my queries about religion to the *Guru Granth Sahib* (the sacred book of the Sikhs) and told me that it would answer all my questions. He is content in being a good Sikh, visiting *gurdwaras* regularly and earning a livelihood without cheating others. Since he works hard and does not cheat others, he can eat and sleep well he said. He contrasts his situation with that of rich people who can neither eat nor sleep well. They suffer from diseases and have to take medicine even to sleep well.

Although honesty and goodness are considered virtues, the craftsmen are pragmatic enough to make compromises, especially when they deal with non-personal relationships. Sukhbeer uses three separate sets of bills in the name of three different companies to avoid detection by sales tax officials. He argues that, after all, he is doing business, not running a *dharamsala* (choultry). In business, it is necessary to be clever because everyone expects the other to lie. Therefore, if he does not lie, he cannot survive. Their self image of being moral persons does not prevent them from making permissible compromises in certain contexts, a trait that Madan (1982: 236) notices among the Kashmiri Pandits as well.

The craftsmen view the government also in personal terms. In their opinion, people elect a political leader or a party to run the government but the officials in the government cheat by taking bribes and bringing a bad name to the leader. The leaders have also begun to cheat and make money because they have to recover the huge amounts of money spent on elections. They become cynical when they are told that the government works according to some impersonal rules. 'According to which rule does a sales tax inspector ask for bribes?' they ask. The officials, according to them, can always concoct charges of violation of one rule or another. Hence it is best to bribe them and avoid trouble. They are reluctant to even get loans from banks because that will mean dealing with corrupt officials.

The craftsmen are of the view that a good political party or leader should work for the people. Honest persons who do social service have to be elected. Political ideologies are not considered important. One of them, formerly an active worker of a communist trade union, favours communism but does not like its atheism. Under communism everyone will get enough to eat and a house to live in. He voted for the party that the Akalis supported in the last elections because a decision to that effect had been taken by the Ramgariah Association, of which he is a member.

It can be said from the foregoing discussion that the craftsmen are persons who view themselves and the world around them in terms of a moral code. Being moral does not come in the way of their making compromises in certain contexts, especially in non-personal contexts.

They apply their moralistic world-view to interpersonal relationships as well. Concepts of honour and respect inform their interpersonal relationships, which tend to be intimate, informal and stable especially when they are hierarchical. They are at home establishing relationships across caste and religion, provided they are expressed primarily in patron-client terms. God is also a patron for them and is viewed in the idiom of reciprocity that guides interpersonal relationships. The same idiom of personal reciprocity informs their view of politics as well.

The craftsmen's world-view is also reflected in their entrepreneurial style. Their aversion to 'paper work' comes in the way of grappling with complex accounting problems. They cannot grasp calculations and rationality exercises in which the relationship between means and ends becomes indirect. When a particular rate for some labour job is quoted, they cannot work out whether by taking into account all the indirect costs, it is worth their while to accept the job. Most craftsmen in the initial stages are eager to accept labour jobs even at low rates in the hope of establishing themselves. The residue of income over expenditure, if any, may turn out to be the wages for their labour, but they suffer from a profit illusion—they regard their earnings as their profit. In managing the workshop, they go by their experience and by certain rules of thumb. By trial and error, they can recognize that certain jobs yield more than others. A few of them have learnt about modern management concepts which can help them assess the performance of the workshop. The most frequently quoted concept in this regard is the annual turnover to capital ratio. They say that if the ratio is equal to one, the workshop can be regarded as having broken even. If the ratio increases, the performance can be gauged to be good. Their notions, however, of what constitutes capital, turnover or profit are vague and imprecise. When an old machine is bought and repaired, the amount spent on the purchase and repair of the machine is regarded as capital, but they do not include in their calculations the value of the time they spend on repair. They make a rough distinction between working capital and fixed capital. The cash they carry around with them daily is often referred to as working capital, whereas in another context the term may refer to their wage bill. Profit is regarded simply as the residue of earnings over expenditure. Some include expenditure on the purchase of old hand tools, rent, if they pay any, and such other sundry expenses in their cost calculations, whereas others exclude them. Those who use their own shed or house do not include the equivalent rental value in cost. The interpretations they give to modern management concepts is so *ad hoc* and varied that it is not possible to say that they serve as objective indicators to assess performance.

The craftsmen display a distinctive rationality in conducting their business. Whatever amount is not consumed from their earnings should be ploughed back into the workshop. It is, therefore, a wise policy to consume as little as possible. Applying the same logic, they consider the expenditure that many industrialists incur on entertainment as reckless, if not immoral. One can spend lavishly like that only when one has made enough money, they say. To consider such expenditure as an item of cost is perverse, as far as they are concerned. After all, the money was not really spent on production. It is clear that the craftsmen cannot take account of various indirect costs in their calculations.

The craftsmen have definite ideas about their future lines of investment. Information about profitable lines is obtained mostly through friends—they do not make a systematic investigation as to the feasibility of the project. Their planning resembles wish-fulfilment exercises because the amount of investment required, the way the money is to be raised, the time duration required to implement the project and such other details are not fully worked out. For example, Chander Prakash has heard from friends that the production of bicycle chains or sinkers (a component of a textile machine) is likely to be profitable. He is inclined to produce sinkers because they are being imported at present. He says he has the technical competence to produce them, and is certain that he will make a profit although he has not worked out the expenditure that he is likely to incur on its development. He does not have a time-frame for the project—he is waiting to get married first and his marriage is held up because he has to save money for it.

The craftsmen reveal the same approach to the future of their children. They realize the value of education, indeed of English education, for their sons. They entertain the ambition of having a commerce graduate or an engineer in the family who can help in the family business. Hence, some of them have made it a point to send their children to 'convent schools', by which they mean English-medium private schools. They say that they are prepared to finance their sons' education for as long as they want to study. They have little time, however, to supervize their children's studies and to take steps to realize their ambitions. If a son does not measure up to his father's expectations, he can always join his father's workshop. Systematic planning by working out all the possibilities, identifying the means for a given end, and working out a time schedule to attain it, is foreign to the craftsmen.

From the foregoing description of the craftsmen's world-view, it can be inferred that they possess *ad hoc* rationality. The term *ad hoc* rationality is used in the Weberian[15] sense to refer to a rationality which deals with *ad hoc* exigencies, interests and tensions. It is contextual and compartmentalized rationality; considerations that apply to one context differ from those of another. The way the craftsmen construct their social world in terms of 'honour' also goes along with *ad hoc* rationality. After all, the craftsmen relate themselves to others depending on the particularistic features of those with whom they interact. Similarly, the ease with which they contextualize morality and allow compromises, and their conception of god as immanent and personal are also consistent with *ad hoc* rationality. What is missing is a reflective rationality which systematically relates ends to means, which takes an integral view of man, god and society, seeking consistency between thought and action and across diverse contexts of life. Reflective rationality is also opposed to magic because the latter involves the invocation of divine or supernatural forces to solve *ad hoc* tensions. Reflective rationality goes along with religion, which views the transcendence of existential conditions as the ultimate goal of human existence.

THE SUPERVISORY ENTREPRENEURS

It has been mentioned earlier that this category of entrepreneurs consists of the former industrial workers as well. The world-view of even the latter category of supervisors, however, differs in some significant respects from that of the craftsmen. It is this critical difference that accounts for the relative success of supervisors.

Like craftsmen, supervisors are also comfortable operating in a world of personal relationships with patron-client overtones. This is especially true of the former skilled workers among the supervisors. They, however, realize—some from bitter experience—that the patron-client idiom does not always work and that they have to learn to

[15] It is evident that my definition of *ad hoc* and reflective rationalities owes much to the terminology Weber (1971) uses in his essays on the sociology of religion.

deal with formalized market and bureaucratic relationships. Further, they learn to discriminate the contexts in which the patron-client idiom is effective from those in which it is not so.

The distinctive feature of the supervisors, as compared to the craftsmen, is their ability to learn to cope with formal relationships. Here the case of Paul, a former skilled worker and a highly respected *ustad* in the township, is pertinent. He groomed a small boy who had come to him in search of work into a highly skilled worker. Gradually he began delegating the supervision of work to this worker and even entrusted the management of his finances to him. As Paul himself said, he liked him more than his own son, and had even taken an insurance policy in his name. He knew that his other workers resented the special treatment given to the boy but ignored it. In spite of this special treatment, Paul's protege left him when he got a secure job in a large factory in Faridabad. Paul now says that you cannot expect gratitude from workers any longer, hence they should be treated as just labourers. Ahuja's case highlights the increasing formalization of work relationships. Ahuja, a former skilled worker, employs 50 workers although he shows only 18 on his records, to avoid the application of the Provident Fund Act. As he is particular that his expensive machines are fully used, he closely supervizes work, and deliberately avoids intimacy with his workers lest they become indisciplined. He justifies his attitude by saying that, after all, having paid so much to his workers, he has to get his money's worth back from them. This does not, however, prevent him from being more flexible with his experienced workers as he realizes that they cannot be easily replaced.

The instrumental orientation the supervisors display towards their labourers is brought to bear on other relationships as well. The younger supervisors learn to emulate some of the executive entrepreneurs, and actively seek to cultivate people in important positions in private industrial firms by giving them gifts and entertaining them on important occasions. The older supervisors, however, show a marked reluctance to deal with government personnel. They are even prepared to forego the various benefits and concessions that the government offers if it means avoiding government officials. Nagi, a Ramgariah supervisor, knows that he is eligible for quotas of steel, as many illicit brokers have offered to help him get them for a commission. He is not willing to do so as he feels that if he accepts the quota, he will be even more vulnerable to harassment by government personnel.

Those who need special raw materials on a regular basis, however, are forced to deal with the government.

The older supervisors, unlike the craftsmen, do not feel vulnerable in dealing with government personnel. They know that the latter visit workshops not to discharge their duties but to collect bribes. They are prepared to accept this as a necessary evil as long as it is within what they regard as reasonable limits. If they feel that the limits have been crossed, they prefer to hire lawyers and fight out the case in a court of law rather than succumb to the harassment. Thus, Paul narrated the instance of some sales tax officials trying to sell him tickets for a charity film show in aid of war widows which had taken place six months earlier. Paul refused to pay Rs 500 for the tickets that they were demanding. As he expected, his refusal resulted in his being booked for some violation of the sales tax law. He hired a lawyer and fought the issue in court. Although he lost the case and had to pay a penalty, he is happy that he put the officials in their place.

While government personnel can be avoided to a certain extent, or kept at a distance, it is necessary to come out of one's shell and actively cultivate relationships with influential people like purchase officers in private industrial firms to ensure a steady flow of job works. Paul, for instance, celebrates the anniversary of his workshop in great style and invites many people whom he considers important for the occasion. As a former foreman in a leading tractor manufacturing corporation, he had met two apprentice engineers who had been assigned to him for training. Paul maintained his friendship with them and used their personal contacts in Bombay to procure job works. To win friends and influence people, some of the older and established supervisors have joined the local branch of the Lions Club while many others have joined several religious associations.[16]

Cultivating networks may reveal an instrumental orientation, but the supervisors mostly use the idiom of patron-client relationships to articulate them. Network articulation also involves the use of caste and kinship, religion, region, language and in fact any bond or loyalty which can establish a personal relationship based on diffused reciprocity. Though reciprocity is involved, the relationship sought to be established is rarely based on the principle of equality. Being products

[16] The Faridabad industrial entrepreneurs are, in general, not too keen to cultivate links with politicians. This is because the dominant local politicians are rural based, and are at best indifferent to the industrialists.

of a hierarchical milieu, when they articulate networks they only reinforce hierarchical values. Like the craftsmen, the supervisors also view society using the notion of 'honour'. Their aspiration to be leading industrialists is related to their evaluation of industrial entrepreneurship as highly prestigious.

For the supervisors, using and articulating networks can be undertaken within a value range. The former skilled workers among them share the craftsmen's perception about 'eating and drinking'. Even the younger supervisors who are more receptive to the idea of cultivating networks are of the view that if one solely depends on network articulation, one will be no different from black marketeers and smugglers. They are keen to succeed financially but only as industrial entrepreneurs.

It is clear from the foregoing that there is a considerable overlap between the supervisors' and craftsmen's world-views. Yet, the supervisors' world-view is distinct from that of the craftsmen in some critical ways. For instance, the supervisors show a better capacity to transcend a world-view based on interpersonal relationships. They differentiate the government from the officials and politicians who are associated with it. They regard the latter as working for their selfish ends, thereby defeating the very purpose of a democratic government. The government is meant to look after the welfare of the entire society. The government may often have to adopt policies which may adversely affect one or another section of society. They regard this as justifiable if the people who run the government are impartial and possess integrity. Only when the politicians are honest and genuinely serve the people, is it possible for our country to progress like America and Russia, they say. They regard the present state of affairs to be so bad that everyone has become corrupt. Corrupt politicians and government officials force others to become corrupt as well. In such a situation, one's survival is itself dependent on the extent to which one pursues one's self-interests.

The supervisors' capacity for the use of abstract concepts is revealed in the manner in which they tackle the complexity of management that accompanies growth and expansion. The supervisors, even the former skilled workers among them, learn the value of 'paper work' and become more receptive to the use of concepts of modern management. They estimate cost, taking account of not only expenditure on raw materials, wage bill, electricity charges and the like, but also of expenditure on 'overheads', such as the amount spent on developing jigs and fixtures and other special tools required for the manufacture of an item. They also make use of standard measures such as the 'profit to capital ratio', 'annual turnover to investment ratio' and the 'overheads to total cost ratio' to monitor the firm's performance. In order to save on income tax, they follow accounting procedures in which the firm is differentiated from the person of the entrepreneur, who is shown as drawing a monthly salary for services rendered. In the case of the more prosperous firms, the entrepreneur is also paid expenses for car maintenance, a house rent allowance, insurance premia and other such perquisites. To avoid or reduce income tax, expenditure on entertainment and gifts are shown against the firm's account as cost. In using these measures, the supervisors hire accountants to maintain the books and occasionally to help argue one's case with the tax officials.

Although the supervisors use concepts of modern management to manage their firms, they do not fully accept the implications of the use of such concepts. Thus, although, according to the modern management practice, expenditure on entertainment is treated as part of their costs, they do not easily reconcile themselves to such an outlook. Notwithstanding what the tax consultants say, they regard expenditure on entertainment as wasteful and would consider it as being spent out of their own earnings. They try to reduce the actual amount spent on such entertainment, although for tax purposes they may exaggerate the amount in the books kept specifically for the purpose. Similarly, the separation between the firm and the entrepreneur is treated as legal fiction at best. Thus, in the use they make of some of the modern management concepts, the supervisors reveal a tongue-in-cheek attitude. They adopt the practice of keeping two separate sets of accounts—one a secret account for their own use and another to satisfy various government departments.

The supervisors, therefore, can develop to a greater extent than the craftsmen. However, they are also likely to hit a threshold when they reach an investment level of around Rs 7 lakh. Although the supervisors use several concepts and indicators of modern management, they use them as rules of thumb rather than as a battery of interrelated indicators which together may point to systemic strengths and defects within the firm. Similarly, the supervisors also draw up 'plans' for expansion and diversification, leaving many factors to chance. In fact, cases of expansion or diversification when analysed reveal the prominent role of fortuitous circumstances. Thus Ahuja undertook the expansion of the firm when, much to his surprise, an application he had made out several years ago under the hire-purchase scheme of the National Small Industries Corporation (NSIC) materialized. He not only obtained an imported lathe and financial help from NSIC, but was also allotted an industrial shed in the newly developed industrial area in Faridabad on the strength of NSIC assistance. Amir Chand's expansion was based on the accidental fact that he heard one day that someone in the Faridabad area was selling a workshop capable of manufacturing u-springs for automobiles. He managed to persuade some of his friends to join him as partners and raised the amount of Rs 6 lakh for the purchase. The supervisors, like the craftsmen, do not systematically analyse alternative investment possibilities.

Another important reason for the operation of a threshold among the supervisors is their inability to delegate responsibility to professional managers and control them. The supervisors generally prefer to keep critical decision-making positions within the family, amongst people whom they can trust completely. With growth, they find it advantageous to convert their firms from a private proprietorship or a partnership concern into a private limited company. Interestingly, one general phenomenon in Faridabad is the high casualty of partnerships over time, unless the partners happen to be related through kinship or marriage.[17] As the firm becomes a private limited company, it becomes more public and open to scrutiny by the government. Along with this, the sheer complexity of management increases. The supervisors find that the old system of informal control is no longer sufficient. They need to decentralize the administration and find it imperative to appoint professional managers. The supervisors are not comfortable dealing with professional managers. They hold that an 'outsider', however competent, cannot appreciate the risks involved when he takes critical decisions. According to the supervisors, 'after all, a hired manager is not playing with his own money'. Besides, they cannot be trusted to keep the firm's secrets. Further, having worked under professional managers in the past, they find it difficult now to delegate work to them and control them. They cannot easily adjust to such a reversal of roles, indicating how hierarchical values

[17] Paul is an exception. His partnership with a former workmate is extremely stable. Paul is the dominant partner, and the mutual trust and friendship extends to their wives also.

constrain and cramp the supervisors' entrepreneurial style. Their social background being mainly lower middle class, the supervisors cannot also find people among their kin who are qualified managers. Thus, the growth of the firm forces the supervisors to operate in a universalistic idiom. Having been used to a parochial setting, they now find it difficult to adapt themselves to the new situation.

The supervisors' world-view is also dominated by *ad hoc* rationality. They tackle problems as they come rather than anticipate problems, methodically work out a strategy for solving them or implement the strategy within a time frame. This is partly due to their preoccupation with day-to-day problems.

The supervisors, with the lone exception of one young man, are all religious. Although there are some Sikhs and one Muslim among them, the supervisors' mode of interpretation of religion is broadly the same. They consider themselves, like the craftsmen, to be moral beings. Their morals also value goodness, honesty in dealing with others, and respect for elders. The supervisors, like the craftsmen, regard religion as overarching, affecting both the work and non-work worlds. There is also an emphasis on pragmatic compromise. The lone Muslim supervisor, Mohammad Saeed, realizes that he cannot devoutly observe many of the rules of Islam, like doing *namaz* five times a day or observing the stricture against taking interest. Saeed says that the Islamic prohibition against interest is really against usury. Regarding the observance of *namaz*, he says that Islam ranks proper duty to other human beings as higher than duty to Allah. If one faithfully conducts oneself in society, Allah, according to him, may forgive one for failing to do one's duty to god. He has displayed a sheet of decorated calligraphy in Arabic in his workshop, which is an excerpt from the Koran praising the virtues of Allah. He has also allowed his workers to put up lithographs of the gods Vishwakarma and Hanuman. He has rationalized his compromise with idol worship by saying that the Koran mentions that when a Muslim comes across people who believe in other gods, he should hold firmly to his own faith and let the others follow theirs.

The supervisors not only make compromises with their religious beliefs but also seek to provide rationalizations for their deviations from the prescribed ideal conduct. These rationalizations reinterpret religion to suit their respective life-worlds. This could be seen in the case of Mohammad Saeed. Another example is that of Sahni, a senior supervisor with a reputation of being a large manufacturer of shoemaking machines, who claims to be religious without being ritualistic. He regards rituals as meaningless, and claims that he observes religious rites only when his wife prevails on him to do so. He says that he believes in *vedanta* philosophy which teaches one to do one's best in this world. In Sahni's version of *vedanta*, one should strive for the higher goals in life. He is waiting for his son to finish his engineering course to take over so that he could relinquish his entrepreneurial work and devote himself to religious pursuits.

Religion, besides guiding action in the world and legitimizing a life-style that may deviate from the prescribed path, is also regarded as offering protection in the mundane world. The supervisors also look upon god as a protector who may punish men or put them on trial. Visiting temples, *gurdwaras* or mosques to offer worship or going on pilgrimage are common among the supervisors. Uttam, a former communist trade union leader, has now become a staunch devotee of Hanuman. He believes that he had to face many problems in his workshop and his home because he had neglected visiting the Hanuman temple every week. He says that ever since he resumed his temple visits, his problems have sorted themselves out. Paul is a devotee of a guru called Charan Singhji Maharaj of Beas. He holds *bhajans* in his house every week and contributes one-fifth of his earnings to his guru. Paul believes that his guru has been protecting him from harm. Nagi, a Ramgariah Sikh, holds *akhand path* in his house regularly and visits *gurdwaras*. He always goes to a *gurdwara* to pray whenever he has a problem. He says that if you please god by hard work, god will never let you down. Nagi is particularly worried about the new trend among the younger generation of Sikhs of shaving off their beard and smoking and drinking, and fears that god will punish them one day. His solution to all the world's ills is simple: if everybody is religious, there will be no cheating and no fighting in this world. Everyone will be protected by his respective gods and there will be no unhappiness.

It is appropriate to mention here the lone case of Banga, who claimed to be modern, scientific and hence an atheist. A 35 year old bachelor, Banga lives with his brother who is working as an officer in a Faridabad factory. Banga says that all religion is superstition. Despite his radicalism, he is no exception to the rule—he keeps lithographs of Vishwakarma and a plaster of paris bust of Shiva and Parvati on a shelf in his workshop. He has decorated the images of gods with small electric lights and has even placed an electric imitation of a *diya* or traditional lamp in front of his gods. He explains this aberration by saying that it is necessary to respect his elder brother's wishes and that of his workers. Banga's atheism, like the religiosity of other supervisors and workers, is *ad hoc* because he recognizes that it is relevant only in a given context.

The supervisors' religions are also dominated by *ad hoc* rationality. When the supervisors call upon god directly or through a guru to intervene in their affairs on their behalf, they recognize that this can be done on numerous occasions and in numerous ways. The path they choose, be it prayer, worship or ritual, depends on the particularities of the context in which they are located. A reflective rationality, based on the recognition of a higher goal and its attainment through a methodical and systematic pursuit of appropriate means, is therefore absent.

It may be noticed that the supervisors view god as ubiquitous. This is true of both Mohammad Saeed and Sahni. Mohammad Saeed sees Allah as watching us all, as being everywhere, all powerful and as the embodiment of compassion. Whenever he encounters difficulties, he undertakes a pilgrimage to Ajmer to pray at the famous Muslim shrine there. Even Sahni, who also stressed transcendence, is not averse to viewing god as someone who can be called upon to intervene in the affairs of men.

To describe the supervisors' interpretations of religion as expressing *ad hoc* rationality is a true, though partial, understanding of their world-view. The attempts they make at reinterpreting their beliefs reveal a tendency towards reflection, abstraction and rationalization. Nevertheless, these reinterpretations do not become consistent and overarching systems. The supervisors hit a threshold because they cannot trust 'outsiders' with critical decision-making powers. This threshold is as much a social constraint as it is mental. It should be pointed out that this threshold is not as constraining or as difficult to overcome as the corresponding threshold is for the craftsmen. In fact some of the younger supervisors who are not only better educated but are also striving to learn more about modern management by subscribing to various journals on management and by enlisting themselves for short term management courses may, in the course of time, cross it.

THE EXECUTIVE ENTREPRENEURS

The executive entrepreneurs possess training as executives and managers in industrial firms or in the government. There are 10 executives who are trained engineers. The remaining six are former senior officers and administrators in the government, army, navy and private industry. The latter group find themselves at a slight disadvantage in running their enterprises but it is a handicap with which they can cope.

The executives, as mentioned earlier, belong to the middle and upper middle classes, but are mostly from upper caste families. This category consists of entrepreneurs from Maharashtra and West Bengal, apart from others belonging to nearby states. These entrepreneurs are best adapted, along with the industrial leaders, to the industrial and social milieu of Faridabad. They suffer from no handicap and there exists no threshold to their entrepreneurial growth. This, however, does not imply that all the executives always succeed.

The executives' entrepreneurial style reveals a tendency towards reflective rationality. The term reflective rationality refers to a rationality which takes an integrated view of a given problem. In attaining specific goals, the direct and indirect consequences of all available alternative lines of action are systematically evaluated. This contrasts with *ad hoc* rationality, which is based on rules specific to a particular context. *Ad hoc* rationality is contextual rationality whereas reflective rationality is universal because it looks at the connections between diverse contexts as well.

The executives do not face thresholds to growth. They adopt scientific management techniques, but not blindly. They know that it is necessary to make some modifications in the concepts of management to suit their specific needs—such modifications are possible for them because they grasp the theories that inform these concepts. This holds true for their accounting practices, the way they control production and the manner in which they review performance and plan for the future. To give an illustration, Godbole, an executive engineer trained in England, controls his production process, the schedule for accepting new job works and his inventory by solely taking into consideration only two of his most expensive machines. He has worked out that if the two expensive machines were kept busy throughout the day he would have ensured enough earnings to cover the interest on his loans, depreciation, and expenditure on other overheads. He feels his other machines are appendages to his two expensive machines. He says it is better to concentrate on procuring job works for his expensive machines because there are few competitors for such job works and because the rates he receives provide a big profit margin.

Another executive, Rehani, is trained as a metallurgist. He has a foundry and a machine shop. He adds that it is the machine shop that helps to bind his customers to him. The latter would save on transport by getting the castings machined at the foundry itself. Besides, the machine shop provides valuable metal scrap which can be recycled. Foundry work being a risky business, Rehani has found it profitable to accept only bulk orders or, as in the case of iron castings, recurring orders. He talks of the 'learning effect' in the case of iron foundry work. As one gains experience in a particular type of casting, one can reduce the percentage of defective castings. Since iron casting is a highly competitive line, the rates are low, and he finds that he can make a reasonable profit only by reducing the percentage of defective castings, which can happen if he accepts orders on a recurring basis. He accepts single orders for alloy castings, because he can quote his own rate for the job.

To control the work process, the executives undertake periodic reviews of the firm's performance. Some, like Rehani, do it once a fortnight whereas others may do it once a week. Such periodic reviews help them identify weaknesses and rectify them before it is too late. In undertaking such reviews, the engineers among the executives are at a distinct advantage. The others usually consult their trusted skilled workers. They gradually learn some of the rudiments of engineering but they continue to require technical help. As these entrepreneurs grow, they tend to hire professional engineers to help them design new products, undertake systematic reviews and prepare detailed project reports.

Among the executives, Vasudeva proves to be an exception. He does not show the same flexibility and ability to improvize by using techniques of modern management. He once prepared detailed job cards for each worker and wanted them to fill in the card every time they completed a job, which caused much resentment. They complained that it takes time to fill in the card which, in turn, lowers productivity, and that work norms that prevail in a factory cannot be applied without providing the supportive services that a factory provides. Since the workers are dissatisfied, there is a high labour turnover. Vasudeva's expensive machines are not properly used as a result and, because his costs soar, he always has to quote high rates for job works. The high rates put off many prospective customers. Vasudeva, however, is unmindful of the problem, and justifies himself by saying that Indian customers want cheap material and are not prepared to pay the high price required for quality work.

There is another executive, Popli, who is an exception in a different way. He has joined his elder engineer brother and has established a big workshop in Faridabad. The workshop is notionally divided into two sections, each section belonging to a separate private limited company. The effective control of both the firms rests with Popli and his brother. Popli says that he has established the workshop mainly to make money by selling quotas and licences for scarce raw materials in the black market. As he possesses expensive HMT machines, he is getting job works even without trying.[18]

The executives are meticulous planners. Before they enter a particular manufacturing line, they obtain all the relevant information using market surveys, feasibility studies, as well as informal advice from bankers and other industrialists. Their previous work experience in other industrial firms gives them an insider's view and helps them establish contacts with important people. As they grow, they prefer to invest in another firm in the small industries' sector rather than expand the original firm. Only those few who have invested, or intend to invest in lines in which they can reap economies of scale, prefer to expand the firm beyond the limits of small-scale industry.

While planning, the executives are cautious about timing. They know that a project, which may be profitable now, may suffer losses if delayed. They are also aware that in the Faridabad milieu it is not possible to adhere to a strict time schedule. Thus, Kashyap, who is a reputed job work vendor in the township, has had to stall his plan to manufacture bench vices.[19] He has had to postpone execution of the project because of the change in the credit policy of the Reserve Bank of India.

The executives realize that a plan for expansion or diversification may take an inordinately long time to be implemented because of the uncertainties they face. The commercial banks take their own time to

[18] This attitude towards entrepreneurship was rare in my selection of entrepreneurs because I interviewed them in their workshops and factory premises. Those who had established enterprises only in name got eliminated. On the other hand, craftsmen were caught in my research net because I followed the social networks of the entrepreneurs themselves to obtain my sample.

[19] Bench vices, hand tools and bicycle components are considered extremely profitable lines in Faridabad as the entrepreneurs feel that there is vast export potential for these items.

process loan applications, and government agencies like the NSIC often take two years to decide on an application. To avoid such delays, they apply for assistance whenever a special scheme is announced by a government or semi-government organization, irrespective of their immediate needs. By doing this, they ensure that when they are in a position to expand or diversify, they are not bogged down by lack of funds or facilities. In fact, a fortuitous factor like a surprise announcement of a concessional facility by the government has pushed many into actually implementing their own plans for diversification or expansion. Thus Bali, who manufactures bright bars in his workshop, had been wanting to produce bicycle chains. He had already established contacts with bicycle manufacturers, who had expressed keen interest in buying chains. He had even talked to some bank managers who were willing to finance his project, but was hesitating to take the plunge because he did not like to borrow from the bank. Things changed when, one day, an application he had made out in the name of his younger brother for an industrial shed bore fruit. The shed was being sold by the municipality at a concessional rate, which could be paid for in easy annual instalments. He then borrowed from a bank to pay the first instalment and to buy machines to manufacture cycle chains.

There are other examples. Chona had a similar experience. He diversified into the production of tractor parts when his elder brother, who retired as an influential government officer, also got an allotment of an industrial shed in the new industrial area. Thus, although the executives are meticulous planners, they also rely on chance factors to convert their plans into reality. Not being able to forecast when and how the various government agencies make decisions, the executives have to be adept in making *ad hoc* decisions.

The executives combine their *ad hocism* with a reflective rationality. They are acutely conscious of the power of modern management techniques to go beyond superficial relationships, and become adept at adapting them to one's particular circumstances. Further, the executives' rationality is not a purely instrumental rationality. The executives are not narrow-minded rationalists with the sole consideration of maximizing profit. Several of them sacrificed lucrative careers as professional managers in order to become industrial entrepreneurs. A few of them even rejected attractive job offers from abroad. They prefer to be 'one's own boss' rather than work for others. They also aspire to use their knowledge and experience to accomplish a challenging task.[20] Although they tend to be conservative in the initial stages, as they establish themselves as entrepreneurs they prefer to branch out into new areas. This is especially the case with engineer entrepreneurs. They are constantly striving to innovate and keep abreast with latest research by subscribing to various technical journals and magazines. For instance, Anand, an engineer entrepreneur, was not satisfied with ordinary job works. He wanted to manufacture an item which he could sell directly to various large industries. Ultimately, after experimenting with some items like solonoid valves, strainers and barrel pumps, he settled down to producing degreasers and vacuum-forming machines. He now produces these against advance orders from big industries. Anand claims that he is the only one in north India to produce these items. He read about degreasers in an engineering journal and experimented with them to make many improvements on the original design. As a result, he even won a contract with the central government's rocket launching station at Thumba in Kerala.

The executives distinguish themselves from the supervisors and craftsmen in being excellent networkers.[21] The supervisors are also good networkers but they are not as capable as the executives in using modern interest-based associations and other types of professional relationships to extend and cultivate networks. Many executives have even managed to save themselves from ruin because of their ability to articulate networks. Vasudeva, for instance, confessed that in the initial phase of his entrepreneurship he had suffered heavy losses because he could not get reliable workers. Matters came to such a pass that he even contemplated selling his workshop. At that time, he was rescued by a friend working as a general manager in a leading public sector undertaking in Calcutta. This friend used his authority and influence to get Vasudeva lucrative jobs in which the vendee firm also supplied the imported alloy, not easily available, required for the job. The profit margin for the job was so high that it helped him avert what would have been a disaster. Another executive, Bhatnagar, also had a similar experience. Bhatnagar's problems began when his vendee, a large industrial unit in Faridabad who owed him a huge bill, suddenly folded up. The collapse brought Bhatnagar to the verge

[20] The executives come close to the description of achievement-oriented entrepreneurs in the McClellandian sense (see McClelland and Winter 1969).

[21] Networkers are those who cultivate and use social networks. See Saberwal (1976).

of bankruptcy, but he was saved by a former classmate who had been transferred to Faridabad at that time as the manager of a branch of a nationalized bank. Bhatnagar's chance meeting with his friend resulted in his obtaining a big bank loan which bailed him out.

The executives are highly conscious of the value of 'right contacts' or 'approach'. They realize that in a government organization or private industry, one way of getting things done is by converting an essentially impersonal situation into a personal one. This can be done if some key person in the organization is 'known' to them, say as a friend's friend, a distant kin or affine, a former classmate or a former colleague. The executives use the patronage of such a person to influence the decisions of the organization in their favour. They actively build and articulate such social networks, using both primordial and modern identities. This may involve doing a favour to an entrepreneur belonging to one's own caste or religion by putting in a good word on his behalf to a friend who is a purchase officer in a leading industrial firm. It may also involve doing a favour to a former colleague in the army by locating a suitable industrial shed for the latter's use. In both cases, what is sought to be done is that a personal relationship is established based on the norms of diffused reciprocity, using primordial loyalties or professional ties. Networking is not only a long-term investment in people but also adds to their stature. The executives seek to extend networks both up and down the bureaucratic hierarchy.

To extend networks, the executives become members of religious and caste-based associations, interest-based associations, professional associations and recreational clubs. A few examples of caste and religious associations are the Sanatan Dharam Zila Samiti, the local branch of the Arya Samaj, the Jain Yuvak Mandal and the Ramgariah Association. The Faridabad Industries Association (FIA), the Delhi Management Association, the National Productivity Council, the Confederation of All India Bright Bar Manufacturers and the Institute of Indian Foundrymen are some examples of interest-based and professional associations. They also have a choice of becoming members of the Lions Club, Rotary Club, the Country Golf Club, the Delhi Gymkhana and so on. Membership of such associations and clubs bring a person in contact with important people in several fields, while interest-based and professional associations protect and promote collective interests.

As members of these associations, it is an advantage to possess some public relations and organizational skills. Such skills make the executives popular and extend their networks further. Members who possess such skills become the spokesmen of the association concerned and, as such, become influential persons in society.

Membership of associations has its costs as well. Extroversion of this variety may adversely affect entrepreneurial achievement. The executives, especially the engineers, avoid becoming members of such clubs and associations till they get established as entrepreneurs. They make an exception in regard to certain professional associations because their journals and meetings provide the latest trends of research in their respective fields. These entrepreneurs also recognize the value of networks and use them when needed. They are keener on establishing contacts with lower level officials—like purchase officers and members of the clerical staff—people with whom they have to transact business on a day-to-day basis.

Networking among the executives has its own culture. One has to attend club meetings and formal parties organized in honour of important officials, bank managers, representatives of foreign corporations and others visiting the town. The parties are organized on a lavish scale. In such parties alcoholic drinks may be served and there is often Western-style dancing. The executives, especially the senior ones, attend such parties frequently. As they belong to middle class and upper middle class urban families, they take to such parties with ease. Even teetotallers accept the necessity of such parties for business purposes. It is only when an executive like Pratap Singh is involved will the exclusiveness of this culture clearly emerge.

Pratap Singh, a displaced person from Pakistan, began his career as an ordinary worker when he arrived in India. He attended an evening school in his spare time and worked his way up the social ladder by hard work. Over the years he obtained a diploma in engineering from a London institute through a correspondence course. He gradually rose in the large firm in which he was employed to become a works manager and a close confidant of the managing director. When he found that his managing director betrayed him by bringing in an officer above him he quit the firm to establish his own workshop. As he progressed in his career he had to consciously learn the etiquette of cocktail culture. When he was a manager he joined the local Lions Club. He proudly mentions that he has maintained a cent per cent attendance record at the club meetings for several years. As a member of the club he gradually learnt the art of making speeches in English and of organizing various club activities, including an annual badminton tournament in Faridabad. As he progressed in his career and in his social life, he found his rustic wife a handicap for him. According to him, if he can go to a party, dance there with somebody else's wife and occasionally give her a lift in his car, his wife too should be able to hold her own when talking to other men. He, therefore, arranged for her to receive special coaching in English conversation. He even tried to persuade her to learn Western dancing, but she refused. Pratap Singh's case dramatizes the culture gap that separates the executives from the supervisors and craftsmen.

Networking does not succeed everywhere. In dealing with government bureaucracies, the executives have to pay bribes and networking alone is not sufficient. It may have to be supplemented by bribes to the lower level functionaries. Many executives narrate instances of higher officers having passed orders whose implementation was deliberately delayed by those down below on one pretext or the other, till a bribe was given. They complain that with so many government regulations, licensing and so on, no industrial entrepreneur can remain honest. 'In order to be honest one has to be dishonest', they remark, suggesting that even to contribute productively to the society one has to bribe. Bribing, however, is an art. One should know to whom and how much to pay. If a bribe is offered to the wrong person or at the wrong time, it may lead to disaster. Thus, Anand, for instance, once had to deal with an officer who belonged to his own caste. He was hesitating to bribe him; he was afraid that the officer may consider it an insult that even a member of his *biradari* (fraternity) displayed no confidence in him. And if the officer happened to be a man 'who cares more for money than for his *biradari*', not paying the bribe would not get his work done. He finally used an intermediary to pay the bribe and incurred the wrath of the official. There are now commission agents in Faridabad who get such things done, thereby saving such embarrassment for the entrepreneurs.

Although they accept the regime of corruption as a necessary evil, the executives are not innocent victims of the system like the craftsmen. They are aware of the rules and regulations and are capable of fighting for their rights as individuals. If some officer or clerk harasses them unduly, they approach higher officers and may even write to higher authorities complaining against the erring official. Thus, Chona, whose brother had been allotted a new industrial shed, met several high officials in the government and the Administrator of Faridabad municipality to complain about the harassment he was subjected to by the officials while getting clearance for erecting the new plant. His complaint resulted in the suspension of an officer in the District Industries office, and he obtained the necessary sanctions promptly.

The executives are well equipped to protect their individual interests. They are at home in a world of person-based relationships as well as in impersonal market-like or bureaucratic settings. Their networking ethos regards social linkages as valuable resources, displaying thereby a manipulative attitude to the social world. The executives are pragmatic people who show a willingness to work within the system. They admit their dependence on social linkages—even the more independent-minded and achievement-oriented executives accept that their success depends on the social networks that support them.

The executives' education and social background enable them to objectify themselves and their own situation. That is, they can look at themselves as others view them. This ability of self-objectification is useful in articulating their personal as well as their collective interests in the various public forums and government committees. This ability is exemplified by the representation made by a committee set up by the FIA to argue before a state government committee for the promotion of small and medium entrepreneurs. The committee, which was headed by Vasudeva, prepared a memorandum urging the state government to simplify procedures to obtain loans, lower the percentage of securities and collaterals required for obtaining loans, extend hypothecation facilities to small and medium entrepreneurs, provide common facilities like a heat treatment plant, create a body to provide technical help, and improve procedures for the provision of raw materials at low rates, marketing facilities and the like. The committee asked the government to do virtually everything to the small and medium entrepreneurs to ensure that they succeeded. The committee based its entire list of demands on the plea that by promoting small and medium entrepreneurs the state government will be creating employment opportunities and speeding up the development process. Clearly, Vasudeva and his colleagues on the committee knew how to put across their requests. The memorandum almost suggested that the government will be doing itself a favour by fulfilling the demands of the small and medium entrepreneurs.

The executives extend their capacity for self-objectification on the topic of workers and trade unions as well. About four of them have had to face 'labour trouble' for one reason or another. Suresh Gupta, a young engineer, is still visiting the labour court for terminating the services of a worker who was 'spoiling the discipline' in his workshop. He says that prior to the entry of communists in Faridabad, the situation was better. Now it 'is the era of workers. Capitalists have to adjust themselves to the new era.' Adjustment for him means being strict with the workers right from the time of recruitment. One has to consult a lawyer and keep all records properly. He asks rhetorically that when, in the modern world, one cannot expect gratitude from one's own children, how can one expect the workers to be grateful even if one looks after them like one's own children?

Regarding the role of the government, the executives do mention that legislation to protect labour is necessary. Otherwise, the labourers may not get their dues. What they object to is the encouragement the government is giving to 'labour indiscipline'. Workers are becoming 'too arrogant' and 'lazy', they complain. They also complain that the government treats industrialists as 'worse than criminals'. They attack the government's taxation policy and its tax collection procedures. According to them, the taxation policy and the regime of controls and quotas are such that no honest entrepreneur can survive. Dinesh Jain, who produces ball bearings, describes his everyday experience in this connection. He had to buy some special type of steel in the black market because he could not get sufficient stock of the material through quotas. The moment he bought it and brought it into Faridabad, he had to pay bribes at the octroi office as he had no proper bills. Having paid the black market price and the bribe, both of which he could not show in his books, he had to fudge his accounts suitably to at least make up for the bribes. He says that black money is created by the wrong policies of the government. Quoting the example of a relative of his, he is now afraid for his children. Since he cannot legitimately invest the black money in his possession, his children will one day gain access to it and, not knowing its value, they might squander it and acquire bad habits. Kashyap, a senior and reputed executive entrepreneur in Faridabad, also voiced similar concern and suggested that the government should provide social security benefits for his family according to the income tax he pays. If that is done, he can at least be certain that if he has an accident or suffers ill-health, at least his family is looked after. Kashyap argues that the government should recognize the contribution that the entrepreneurs are making to the society and honour them instead of treating them as 'worse than criminals'.

It should be stressed here that while the executives favour liberalization they do not favour a *laissez faire* ideology. They want the government to act with benevolent paternalism to protect and encourage entrepreneurs, especially small and medium entrepreneurs. They realize that if the government completely withdraws, they may not be able to survive the ravages of free competition.

The life-worlds of the executives are dominated by work, but they do make a sharp distinction between the work world and the non-work world. This mental distinction is sustained by the physical separation between the home and the workshop. Six of the 16 executives interviewed live in Delhi and commute daily to Faridabad. Some of the more prosperous executives maintain two houses—one in Faridabad and the other in Delhi. It is a status symbol to possess a house in Delhi.

The sharp distinction between the work world and the non-work world is necessitated, to a certain extent, by the manner in which the executives organize their work time. They are creatures who regulate their life by keeping a diary to cope with the complexities of their work. To meet important people in industry and in the government, they invariably have to take prior appointment. They also have to meet the deadlines of delivery dates for sending prototype samples and tender quotations, filing sales tax and income tax returns, and so on. All this means that they have to carefully schedule their visits to raw material suppliers, purchase officers of vendee firms, bank officers, accountants and others so that work in the workshop goes on continuously and is not hit by the paucity of raw material or finance. Despite such organization, they have to be prepared for emergencies in the workshop, surprise visits by factory inspectors, sales tax officers and others. Other unanticipated problems like unscheduled load shedding of electricity, inability of a dealer to supply raw material on time, and some delay in the octroi office in getting goods across the Faridabad-Delhi border may arise needing immediate attention. Hence, they always seem to be running behind schedule and are continuously revising their schedule of work. Amidst such hectic activity, they have to carefully schedule their non-work life as well in order to attend to the needs of the members of their households.

The separation of the work and non-work spheres is, to some extent, necessitated by the structure of the executives' households. The 12 married executives live in simple households[22] with their wives and

[22] For a definition of 'simple household' see Shah (1973).

children. Their parents live separately with their unmarried children. The executives maintain intimate ties with their parents, siblings and affines. The arrangement combines the advantages of extended family ties while allowing a certain degree of privacy and autonomy to the executives in matters relating to their nuclear family affairs. The fact that they live in separate households imposes certain household obligations on the executives. The executives who are relatively free from such obligations are the four bachelors.

Although the executives ideally prefer to keep their work concerns away from their non-work world, they do not always succeed. As they cannot exercise much control over their work schedules they have to make inroads into the time meant for non-work world activities. Added to this is the overlap that occurs between the work world and the non-work world. The birthday of an important industrialist or a purchase officer may involve attending his birthday party along with giving an expensive gift. Occasionally, to promote business interests they have to organize expensive dinner parties at home or at a five-star hotel. These parties symbolize their success and creditworthiness. Hence, the executives spend much more money than they can actually afford on such parties. In fact, according to a cynical young executive, the more indigent an entrepreneur is, the more lavishly he spends on such parties.

Apart from the exigencies of work, there are other demands made on their non-work world. Unlike the craftsmen, the executives maintain links with their distant relatives. Such links entail obligations like attending life-cycle ceremonies and other religious functions in their relatives' houses. As the executives know the value of education, they also spend their non-work time supervising their children's studies. Since they are also religious, the time for visiting temples and attending talks on religion will also have to be found. Lulla, a Sindhi executive, makes it a point to visit a Hanuman temple in Delhi every Monday and goes on a pilgrimage to Shirdi, near Bombay, every year. Vasudeva visits the Ramakrishna Mission every week and attends religious discourses there whenever they are organized. Hence, in view of so many demands on their time, they find it imperative to plan rest and recreation as well. Playing bridge on Saturday evening at the club or taking their wives for a film show on Thursday are some of the compulsive rules they adopt to make certain that a private universe of leisure is created.

From the description given earlier, it should not be inferred that the executives regard religion as belonging to the non-work world or as competing with leisure. They also regard religion as central to life and provide elaborate rationalizations to support their positions. Their interpretations of religion, be it of Hinduism or Sikhism, are not just *ad hoc* rationalizations but an entire system of beliefs. It is quite common to come across executives who follow a guru. Their systems of beliefs are the ones handed down by the respective guru. Thus, Bhatnagar, who is a follower of a popular guru from Bombay region, argues that *dharma* is based on the scientific principle of harmony in this universe. Different life forms have different natures, and each one works according to its *dharma* or nature. He even claims that the idea of caste is scientific. In the past, the caste system was enforced because each person had to do that which suited his nature. The system is distorted today because even people who make shoes claim to be Brahmins. But each one has a duty to perform—has to find resonance with the universal harmony. Religion, here, is interpreted as doing one's duty—following one's *dharma*. When in doubt about *dharma* one has to consult one's guru, who clears the confusion.

Pratap Singh, a Sikh entrepreneur, is a follower of another guru hailing from the Bombay region. Though a staunch Sikh, he does not find that fact clashing with his faith in this saint. This guru does not ask anyone to change his religion. All that he preaches is that life is like a vehicle on three legs—the legs of materialism, spiritualism and intellectualism. All the three legs have to develop proportionally if life has to remain balanced. The right approach to life is to develop in all the three directions simultaneously.

In such interpretations, there is an effort on the executives' part to legitimize their life-style. Their belief systems affirm and accept their worldly concerns. Religion is viewed primarily as a set of private beliefs. Rituals are regarded as secondary. They do observe some family rituals but the right attitude of mind is regarded as more important. Rituals are supposed to reinforce such an attitude.

The executives also call upon divine intervention in their worldly affairs to protect them and promote their worldly interests. When they talk about their gurus they refer to their spiritual attainments, and quote instances of miracles performed by them. Some claim that ever since becoming the particular guru's disciple, the guru has invisibly sorted out all their problems. Many of the executives believe in astrology and consult astrologers to take decisions concerning secular matters. As with supervisors and craftsmen, the executives' religion contains a magical element, in the sense that they seek to bring divine forces to aid them in their worldly affairs. Magic co-exists with reflective rationality.

Considering that the executives do not face any thresholds to entrepreneurial growth, it can be concluded that the Faridabad milieu is supportive of entrepreneurs who possess the ability for abstraction and self-objectification, and are capable of reflective rationality. They accept the *status quo* and work within it. This involves, among other things, accepting parochial loyalties and using them. This approach is antithetical to reflective rationality which stresses universalism. The executives become adept at walking on two legs—the leg of reflective, universalist rationality, and the leg of *ad hoc* rationality (that is, magic).

THE INDUSTRIAL LEADERS

The industrial leaders are the patriarchs of Faridabad. Some have an all-India reputation, and are powerful enough to influence government policies. They are on important advisory committees set up by the Central government to promote exports, and are often consulted by the government on industrial policies, labour laws and tax laws. These persons set the standards for others to emulate. Being the leading lights of Faridabad, they lead a public life. Although they consider themselves to be industrialists primarily, their concerns range far beyond industry. They are all respected middle-aged persons now, who are in the process of grooming their sons to take over the reins of corporate control and management. They are highly conscious of the image they project, and tend to be cautious and deliberate in answering questions in interviews. The more prosperous industrial leaders have even appointed media experts as public relations officers in their firms. They also utilize forums like the FIA to do this job for them collectively. The FIA often projects the problems of industrial leaders as those of the entire industry in the region. The world-view they construct contains elements which are consciously meant for mass consumption.

As an illustration of the image they project, a statement from a brochure prepared by one of them is quoted:

> The miraculous achievements of Chanda and Co. are a saga of success unequalled in the area and are exemplary. The high business standards, the discipline and devotion of all those engaged in the

venture *coupled with the business acumen of the Chanda family are asset* [sic] *to which all the success can be ascribed.*

Another industrial leader ascribed his success to his adventurous spirit, uncompromising attitude towards quality and insistence on modern management techniques. He claims that he was able to forge ahead of the others because of his capacity for hard work, spirit of adventure and desire to serve society.

The factors that are repeatedly mentioned as important in accounting for their entrepreneurial success are dedication, hard work, spirit of adventure and business acumen. They speak with nostalgia about their early days of struggle against almost insurmountable odds and the risks they took. In all such accounts, while they may acknowledge the people who provided critical support, there is a predominant stress on their own capacities and achievements.

The industrial leaders are not satisfied with expatiating on their past achievements. They stress the significant contributions they are continuing to make. They advertise their corporations' achievements in glowing terms in newspapers. One such advertisement refers to the support and encouragement the corporation is giving to several thousand small industries through its vendor programme. Another advertisement projects the corporation as a pioneer in harnessing modern technology for the country's agricultural prosperity. A third one mentions the contribution the concerned industrial firm is making for the defence of the country. Although the concerned corporations are public limited companies, they are popularly known as companies belonging to a particular leading industrialist. In popular perception, the distinction between the industrialist and the corporation he heads is inconsequential. An industrial leader is known by his corporation and vice versa. Not only that, an industrial leader himself regards the corporation he heads as if it were his own private estate.

As mentioned earlier, the activities of the industrial leaders are not just confined to the industrial field. They are far too important for that. Two of them have contributed to the building of a hospital with all the modern facilities for cancer research and treatment. They are also active in building a hospital for heart ailments. Some have donated funds to various colleges and engineering institutes, and are chairmen or members of their governing bodies. Others have established charitable trusts in the names of one or the other of their parents for various purposes ranging from scholarships for poor students to special facilities for the education of mentally retarded children. The industrial leaders act as philanthropists.

The industrial leaders are acknowledged as the 'big people' in Faridabad. Although they do not explicitly acknowledge the help they receive from the government, they tacitly seek its patronage. In fact, their advertisements could also be construed as tacit attempts to stake claims for the government's patronage and support. Thus when a leading industrial corporation advertises its vendorship programme, it is also an indirect way of telling the government that support to the corporation is in the public interest. Occasionally, an industrial leader may indirectly acknowledge the help he received from the government. One of them, while recounting his entrepreneurial biography, mentioned how he could enter the area of colour printing and packaging because of the Chief Minister's backing in obtaining an import licence for coloured printing inks. The sole manufacturer of coloured printing inks in the country was a big Marwari industrialist who surreptitiously used to divert a major portion of his output to his subsidiary firms spread all over the country. He compared that Marwari house to 'Hanuman', who could contract or expand in size at will. He apprised the Chief Minister of the prevailing situation and successfully prevailed on him to exert his influence with the Central government to procure the import licence so that local entrepreneurship in the State is not stifled.

To enjoy the benefits of the government's patronage, social linkages with important officers, ministers and politicians have to be cultivated. In the forementioned instance, the industrial leader had managed to gain the Chief Minister's patronage by playing a prominent role in organizing the ruling party's national convention held at Faridabad. The industrial leader organized a reception committee with other leading industrialists of Faridabad to provide hospitality to important party delegates and to make arrangements for transport and other infrastructural facilities. The Chief Minister was thoroughly impressed with the industrial leader's efficiency and, as a gesture of appreciation, granted him the privilege of free entry into the Chief Minister's office. The industrial leader not only got the ink he wanted but was able to get the Faridabad Municipality superceded on the plea that the civic amenities there were in a state of neglect.

Industrial leaders pay a price for political patronage. Occasionally, they have to provide jobs for the relatives of ministers and important politicians in their firms and make generous financial contributions to political parties at the time of elections. Industrial leaders prefer to keep their eggs in different baskets. They contribute to the opposition parties as well to ensure continuance of patronage in case the ruling party gets defeated in the elections. By supporting opposition parties as well, they can prevail on them not to create 'labour trouble'.

Political patronage works only up to a point because politicians have their own interests to safeguard and party politics imposes its own compulsions on public policies. Thus, industrial leaders who are agitated about the worsening labour situation and power shortage in Faridabad, realize only too well that the government cannot afford to support them. They used the FIA to agitate for their demands and to put political pressure on the government. The FIA sent a delegation consisting mostly of small-scale industrialists of the town to present memoranda to the State and Central governments. The FIA also organized a press conference to highlight the plight of Faridabad industries. The FIA spokesman claimed at the press conference that the 100-day old strike in a leading industrial firm in Faridabad had starved hundreds of small industries of their work. The power shortage, he claimed, had brought industrial productivity to a standstill and several thousand workers had been laid off. He, therefore, demanded that the government should take stern measures to put down the strike and divert electricity from domestic consumers to industries. Even industrial leaders know the value of collective political action.

The case of collective action mentioned earlier reveals the public relations strategy of industrial leaders who can couch their interests in such a manner as to make out that the interests of workers and that of small industrialists are threatened.

Industrial leaders take pride in the high quality of their products which they claim is a result of the high standard of efficiency in their plants. They possess expensive and sophisticated machines. A few have established their own research and development organizations with a complement of research scientists and technologists. Their research and development wing systematically explores methods of raising productivity by using time and motion studies, along with systems analysis. The industrial leaders indeed institutionalize reflective rationality with a vengeance.

The recruitment policy of industrial leaders, especially at important levels of decision-making, is based on a combination of particularistic and universalistic criteria. For the top managerial positions they only choose people with proven loyalty and integrity. If there are qualified people available among close relatives or within their own caste or community, they are understandably preferred. Close relatives are sent abroad for higher training so that they can be placed in sensitive and strategic positions in the firm on their return. In Faridabad, 'Madrasis', as south Indians are called, are also in great demand. The south Indians, who are mainly from Kerala, have established a reputation for themselves as being loyal, hardworking and intelligent white-collar workers and managerial personnel. It is easy to discern the recruitment policy of some of the large establishments by even a casual visit to the firm's office. Invariably, an Agarwal industrial leader's corporation will be filled with Agarwal officers and a Sikh industrialists's office will contain a large number of Sikhs. Two industrial leaders espoused a universalist policy in recruitment. They boast that their main criterion for recruitment is whether the person is a 'go-getter' or not. By a 'go-getter' they refer to a person who can get work done on time without giving excuses.

Industrial leaders use particularistic criteria in labour recruitment as well. They find it always advantageous to recruit workers whose good conduct can be vouched for by other trusted workers or officers. They cannot, however, be too particular in recruiting skilled workers because they are not easily available, except that they tend to be wary of active trade unionists. In Faridabad, industrialists informally exchange information about workers who 'create trouble'. Such workers find it difficult to get a job in any of the bigger firms there.

Industrial leaders claim that their workers are their first charge whom they prefer to treat like 'family members'. They encourage their senior workers to set up small workshops and offer to help them with job works. The two progressive industrial leaders have introduced many novel welfare schemes for their workers.

The patriarchal approach of the industrial leaders also emerges in their managerial style. They try to keep crucial decision-making positions within the family to ensure that the critical secrets of the firm are not leaked out. They are now actively grooming their sons to take over from them. While giving considerable autonomy to their sons, they try to keep a tab on the affairs of the firm in many ways in order to ensure that they are consulted while taking crucial decisions. Industrial leaders claim that in decision-making they rely more on their innate business acumen than on organizational inputs. Even when certain lines of action are recommended by senior and trusted managers in the corporation, the industrial leaders may take them by surprise by suggesting options which had not occurred at all to them. For instance, a son of the leading producer of leaf springs presented his father a detailed project report on diversification into the manufacture of moped engines. The project was being sponsored by the Industrial Development Bank of India. The father suggested that instead of manufacturing engines he should concentrate on producing carburettors in collaboration with a Japanese firm. He felt that while even a small industrialist could hope to produce moped engines, the carburettor technology was highly sophisticated and new.

Although the industrial leaders claim to have established independent research and development wings of their own, they prefer to invest in foreign collaboration, even in areas where their research and development wings are active. Foreign collaboration, they say, is like cold war. The prominent industrial leaders mentioned how they advised the government to amend its tax laws to encourage local research. They quote the example of Japan and want the government to support large industries with finance and suitable labour policies so that Indian industry can also make rapid advances.

Industrial leaders project themselves as patrons in the field of religion as well. They have made donations for the construction of temples and the propagation of religion. Being a devotee of a guru has become fashionable. They contribute to the propagation of their guru's interpretation of religion by financing publications carrying the guru's message. Some have been initiated by their respective gurus into meditation which, they claim, has given them inner peace and tranquility. The gurus also solve family problems. Industrial leaders rush to their respective gurus whenever they face a problem—whether financial, health, mental, children's career or marriage. They have faith in the guru's capacity to use his divine energy to sort out their problems for them.

The industrial leaders are actively sponsoring and supporting modern reinterpretations of religion. The reinterpretation is not only of beliefs, but also of rituals. Thus, in establishing a south Indian temple in New Delhi, one industrial leader insisted that all the vedic rituals of consecration be meticulously followed. To that end, he had many of the ritual implements and ingredients collected from different parts of the country. To ensure that only ghee made from cow's milk was used for the vedic ritual, he had cow's milk collected and ghee made in his own house. He invited reputed vedic priests and persuaded the Shankaracharya of Kanchipuram, Tamil Nadu, to preside over the rituals. This is also true of the Jain and Sikh industrial leaders. A Sikh industrial leader, for instance, commissioned a famous Sikh artist to paint a portrait of Guru Nanak. He had prints made and distributed them as gifts to several organizations and private individuals. The painting was also printed in the diary he had specially printed to his specifications, so that copies could be distributed among friends, employees and clients. When he opened a new factory he organized an *ardas* (a thanksgiving prayer) in the factory for which a minister in the Central government was invited. Rituals, far from getting truncated, are observed in all their complexity, especially on public occasions. Organizing elaborate rituals provide a double benefit: it gives religious merit and enhances one's status in the secular world.

Industrial leaders propagate a variety of interpretations of religion, each claiming that his is most in keeping with the true spirit of the scriptures. Some of them even claim their interpretations to be 'scientific'. The interpretations affirm worldly life and worldly success and, at the same time, allow for transcendence. For instance, one popular guru of an industrial leader, who claimed to follow tantrism, advocated that satiation with worldly experience was a qualification for *moksha*—a starving man in a temple thinks of the offerings made to god rather than god himself whereas a man with a full stomach can devote himself to prayer. Hence the industrial leaders are the elect for *moksha*—they can now concentrate on higher ideals. Another industrial leader says that his master assures everyone *moksha* without having to give up worldly concerns. All that is needed is to meditate every morning on the *mantra* into which his master has initiated him. The meditation helps him adopt the right mental attitude to daily life. The leader claims that his master guides him in everything he does: all his success is, therefore, his master's achievements. He regards himself as only an instrument of his master.

To summarize, industrial leaders seek to translate their success into patronage not only in industry, but in other spheres of public life as well. In the field of religion their role parallels that of *yajamana* in ancient India (see Thapar 1982: 275). They also act as patrons in the fields of education, scientific research and health. While utilizing reflective rationality, they actively contribute to the propagation of *ad hoc* rationality and magic.

CONCLUSION

From the foregoing analysis, it can be inferred that the world-view acts as a severe constraint on the craftsmen entrepreneurs and, to a lesser extent, on the supervisory entrepreneurs. In the Faridabad industrial world the craftsmen entrepreneurs cannot develop their enterprise beyond certain investment limits because they are not equipped to comprehend the complexity of the systems within which they operate. They find it difficult to transcend the dyadic world of one-to-one relationships. They even conceive of god in terms of interpersonal relationships. In comparison, the supervisors, especially those who have been educated beyond the high school level, are able to comprehend how the system works, and also cope with it better. To be sure, they cannot handle problems of delegation of authority to professional managers that become necessary as they develop, because of their inability to trust them. In this context, the better educated supervisors are equipped to learn. Hence, the threshold they face is not insurmountable. As regards the executive entrepreneurs and the industrial leaders, their world-views do not present any obstacles to their development.

If we study the distinguishing characteristics of the world-views of entrepreneurs, we find that the supervisors' entrepreneurial style reveals a capacity for abstraction in comparison to that of the craftsmen. This capacity, however, does not amount to reflective rationality—an ability to look at different indices of scientific management as interrelated, which together reveal the specific weaknesses and strengths of the firm. The executives possess this capacity for reflective rationality and can flexibly and imaginatively use the modern concepts of management. They are also capable of self-objectification, which enables them to deal with the external world better. Industrial leaders go one step further by institutionalizing reflective rationality itself.

It is possible to infer that entrepreneurial success in Faridabad requires a combination of reflective and *ad hoc* rationalities. Reflective rationality in itself is not suited to the Faridabad milieu. The styles of functioning of significant others in the Faridabad milieu, including government officers and inspectors, bank officials and purchase officers in the various industrial firms, make it difficult for the entrepreneurs to strictly adhere to their work schedules and plans. The entrepreneurs need to be flexible enough to meet unforeseen contingencies that keep cropping up. The capacity to improvize is an asset. In fact, one executive entrepreneur, who sought to rigidly follow the detailed procedures of modern management, found his problems multiplying. On the same count, *ad hoc* rationality, revealing a capacity for considerable improvization, is sufficient only for a bare and tenuous survival.

The entrepreneurs' way of relating to the world around them differs across categories. The craftsmen can relate to persons across the lines of caste, religion, region and language, provided they can use the patron-client idiom. They avoid meeting government officials and, when confronted by them, as they inevitably are, the only way they can deal with them is by offering bribes. The supervisors can deal with officials better, and they show a tendency to build and extend social networks revealing thereby an instrumental orientation towards social relationships. They are, however, not willing to employ 'outsiders' to manage crucial positions in the firm. Beyond a narrow circle of family and close relatives they cannot 'trust' outsiders in critical positions. The executives and industrial leaders can do this, but they also prefer to recruit closely related kin in critical positions. These two categories of entrepreneurs exhibit a capacity for self-objectification and for dealing with abstractions. The executives and industrial leaders are versatile, and can use both the idiom of hierarchy and the idiom of equality. Hence in Faridabad, a combination of modernity and tradition, rather than modernity or tradition separately, contributes to entrepreneurial growth. This is likely to be the case so long as the Faridabad industrial structure remains hierarchical, with a small but powerful group of industrialists dominating the scene.

Ad hocism informs religion as well. The *ad hoc* view considers the world as consisting of separate compartments.[23] The world is contextualized—different principles and norms are seen as applying in different contexts.[24] Reflective rationality, which strains towards a consistent and universal logic of thought and action as in Protestant Christianity, is itself compartmentalized in the Indian context. New interpretations of religion are being sponsored. These interpretations allow a mixture of magic and observance of elaborate rituals, and allow compromises legitimizing diverse styles of life. The elaborations they offer are reinterpretations of the Hindu scriptures made by their gurus. The latter use their spiritual powers to assure their entrepreneur-devotees protection in this world as well as *moksha* in the next. All this fits in with Agehananda Bharati's account of Hinduism;[25] he holds that

[23] This supports Singer's thesis on compartmentalization (1972: 320–29).

[24] This conclusion runs on lines similar to A. K. Ramanujan's argument in his 'Is There an Indian Way of Thinking?' (1980). I thank Satish Saberwal for bringing this essay to my attention.

[25] Agehananda Bharati (1981: 23–40) says that this has been the case in traditional Hinduism, even of the Sankara school.

transcendence and immanence overlap and are not seen as contradictory in Hinduism. Further, religious principles are themselves seen contextually.

What is true of Hinduism seems to be also true of the interpretations of Jainism, Sikhism and Islam as provided by the entrepreneurs belonging to the respective religions. Although there was only one entrepreneur who professed Islam, his interpretation was also contextual. The Sikh entrepreneurs also interpreted their religion as world-affirming. A devout Sikh can work for his material prosperity as well. In fact, devotion helps in one's spiritual and material advancement, according to many Sikh entrepreneurs. Like Hinduism, Sikhism as well as Islam, as interpreted by the entrepreneurs, could be both world-affirming and transcendental. And, as in Hinduism, compromises are allowed by reinterpreting religious tenets suitably by the respective followers of Sikhism and Islam. In a pluralist social milieu, the followers of each religion tend to look at their own religion as one among the many forms practised. Each religion, even atheism, tends to get contextualized rather than regarded as a universal creed applicable to all contexts. Therefore, the strain towards rationality which Weber found in Protestantism is missing in these modern Indian interpretations of religion.

Turning to ritual, the fact that certain public religious rites get elaborated has to be assessed against the background of the evidence provided by Milton Singer (1972: 332) regarding entrepreneurs' abridgement of rituals in the wake of industrialization. In the context of the elaboration of public rituals by the Faridabad entrepreneurs, it can be inferred that while some domestic rituals get abridged because they have lost their social significance today, public rituals have now acquired a new social and political significance.

Industrialization in Faridabad has, therefore, not destroyed the connection that India's present has with the past. However, continuities should not be exaggerated. A universalist reflective rationality has become an ingredient of success, though it is not sufficient in itself. Contextual, *ad hoc* rationality persists, and is strengthening particularistic ties. Industrialization in modern India is generating mutually contradictory tendencies. While the particularistic criteria of caste, kinship, and so on continue to be significant, a successful industrial entrepreneur also needs to be capable of using universalist norms which belong to a different cognitive universe.

REFERENCES

BERGER, PETER L., BRIGETTE BERGER and HANSFRIED KELLNER. 1973. *The Homeless Mind.* Harmondsworth: Pelican Books.

BHARATI, AGEHANANDA. 1981. *Hindu Views and Ways and the Hindu-Muslim Interface.* Delhi: Munshiram Manoharlal.

HARRISS, JOHN. 1982. 'Character of an Urban Economy: "Small Scale" Production and Labour Markets in Coimbatore (I and II)', *Economic and Political Weekly*, 5 June and 12 June: 945–54 and 993–1002.

MADAN, T. N. 1982. 'The Ideology of the Householder'. In T. N. Madan (Ed.), *Way of Life: King, Householder, Renouncer.* New Delhi: Vikas, pp. 233–50.

MCCLELLAND, DAVID and WINTER, DAVID. 1969. *Motivating Economic Achievement.* New York: Free Press.

PANINI, M. N. 1978. 'Networks and Styles: Industrial Entrepreneurs in Faridabad'. In Satish Saberwal (Ed.), *Process and Institution in Urban India.* Delhi: Vikas, pp. 92–116.

——. 1979. 'A Sociological Study of Entrepreneurs in an Urban Setting', unpublished Ph.D. thesis, Department of Sociology, University of Delhi, Delhi.

RAMANUJAN, A. K. 1980. 'Is There an Indian Way of Thinking?' Paper presented to the First Workshop of the ACLS-SSRC Joint Committee on South Asia sponsored project, 'Person in South Asia', Chicago, 16 September.

SABERWAL, SATISH. 1976. *Mobile Men.* Delhi: Vikas.

SCHUMPETER, JOSEPH A. 1934. *The Theory of Economic Development.* Cambridge: Harvard University Press.

SHAH, A. M. 1973. *The Household Dimension of the Family in India.* Delhi: Orient Longman.

SINGER, MILTON. 1972. *When a Great Tradition Modernises.* New York: Praeger.

SONI, B. M. (Ed.). 1973. *Directory and Who's Who of Faridabad Industries.* Faridabad: Emco Publication.

THAPAR, ROMILA. 1982. 'The Householder and the Renouncer in the Brahminical and Buddhist Traditions'. In T. N. Madan (Ed.), *Way of Life: King, Householder, Renouncer.* New Delhi: Vikas, pp. 273–97.

VAN GENNEP, ARNOLD. 1960. *The Rites of Passage.* London: Routledge and Kegan Paul.

WEBER, MAX. 1971. *The Sociology of Religion.* London: Methuen.

4

A Perspective for the Sociology of Indian Organizations

N. R. SHETH

This essay advocates the need for a sociology of organizations in the Indian context. A search through the appropriate literature has led me to conclude that there has so far been little effort towards developing a comprehensive sociological understanding of Indian organizations, although there is, ironically, considerable awareness and concern regarding the significance of social and cultural forces in influencing organizational structures, processes and effectiveness. Studies in the conventional branches of sociology (such as urban, industrial, educational and political) often include analyses of the socio-cultural aspects of various types of organizations. While these studies enlighten us on some sociological aspects of organizations, they provide a partial and inadequate sociological perspective, as I shall argue on the basis of a brief review of the literature on the subject. I shall examine the theoretical perspectives which have so far guided the sociology of organizations, and then suggest an alternative perspective for a more meaningful understanding of organizational phenomena. Finally, I shall illustrate the issues which, in my view, should be regarded as central to the sociology of organizations.

There are several important reasons why we need to pursue the development of a sociology of organizations within the Indian cultural framework. In the first place, formal organizations are increasingly occupying a central position in contemporary society. Perhaps there is a sense in which organizations can be regarded as the microcosms of modern society just as the village communities were regarded as the microcosms of traditional society. Progressively more and more services—like employment, housing, health, education, recreation and the processing of agricultural inputs and outputs—are rendered to people through formal organizations. Even religious missions are conducted with the help of formal organizations. The use of well-oiled organization machines for clandestine business, which occupies a significant place in contemporary society, is well known. Thus, as Collins (1975: 286) reminds us, 'most of the other things sociologists study—stratification, politics, education, deviance, social change—are based on organizations or take place within them'. The experiences gained by people in terms of their links with organizations—as owners, managers, clients, members, beneficiaries or simply as neighbours—are likely to spill over to the rest of their social existence. The sociology of organizations may therefore constitute a key factor in any attempt to understand the social order encompassing them.

In the background of this emerging importance of organizations, sociological awareness is growing among scholars and practitioners concerned with problems of management and administration. Sociological and social psychological theories on organizations which developed in Western culture (for instance, theories pertaining to human relations, participative management, bureaucracy, and alienation and commitment of employees) have been extensively borrowed by Indian social scientists and management practitioners in their thinking, planning and decision-making. Such theoretical models and perspectives are often used to suit the intellectual or ideological predispositions of the users. Some are concerned with the obstructive influence of Indian institutions and values on the rational designs of organizations, while others are interested in evolving organizational designs to match Indian tradition. Also, the conceptual tools and approaches employed by management experts engaged in organizational interventions (such as organization development, human resource development, and humanization of the work environment) include sociological concepts in the areas of group dynamics, role analysis, culture, socialization and conflict resolution. Such a practical use of sociology in understanding and resolving management problems is likely to attract sociologists to devote more attention to pragmatic research and the quick application of theory and concepts. Consequently, the need for developing an adequate understanding of social behaviour of men and women in organizations may receive low priority in organizational research. The sociology of organizations should provide a more realistic and concrete perspective to practitioners and change-agents to deal with organizational issues.

Another factor conducive to the growth of organizational sociology is the increasing concern among Indian sociologists regarding the role of the sociologist in achieving social goals and successfully implementing social plans and programmes in various spheres of social life. For instance, Srinivas states:

> The sociologist's commitment to democratic processes is fundamental... commitment to democratic processes results in the sociologist having a deep concern in national development. ... Development which makes the rich richer and leaves the conditions of the masses of the poor unchanged is likely to produce chronic political instability... commitment to development is therefore also a commitment to the reduction of economic and social inequalities (1966: 163).

Similarly, Dube (1977: 12) wants Indian sociology to 'address itself to the living concerns of today and tomorrow. For this we shall have to identify critical problems, pose the right questions and devise appropriate procedures of investigation in respect of them.' Mukherjee stresses that

> at this crossroad of its development, sociology in India must have a role of its own to play in order to answer the 'Indian question' in its present context. This role...lies in assuming the responsibility to identify the *soft spots* in the social organism, viz., *those vulnerable regions of the social structure through which change in the society is, or can be, effected* (1973: 49).

Organizations which serve as the dominant carriers of societal aspirations, policies and plans provide the sociologist with an important base for testing the value of sociological theory and concepts in prompting socially desirable plans for change.

THE EXISTING LITERATURE

The current literature on social behaviour in Indian organizations[1] includes studies pertaining to social relations within organizations as well as to social and cultural forces outside organizations which may impinge upon these relations. The following overview of literature covers the relevant contributions made by sociologists and other social scientists (especially social psychologists and management scientists) interested in organizational systems and performance.

Social organizations as units of sociological analysis have at best drawn modest attention from Indian social scientists. Apparently, the scarce intellectual resources in Indian sociology have been employed largely in the study of some traditional institutional and cultural forces (such as caste, extended kinship, village community, peasant culture, ritual, and economic and political segmentation) in their myriad manifestations in rural and urban communities in the country. Over the last two decades or so, however, the proportion of scholars interested in relatively 'modern' social phenomena (such as education, industry, administration, government and health-care institutions) has steadily increased. These scholars deal, in one form or another, with social relations characterizing organizations in their respective fields of interest.

The point of departure for the sociologist's interest in organizations was provided by the assumption that modern industrial, educational and other formal organizations were a product of the industrial culture of the West, which was believed to be incompatible with the traditional culture of India. Hence the conflicting pressures exerted by the organizational goals on the one hand, and the traditional culture on the other, were believed to produce incompatibility in the social behaviour of people within such organizations. Accordingly, the main object of the sociologist's interest in organizations was to examine the ways in which the traditional bonds of caste, kinship, village, agriculture, religion, and so on, were carried over to the formal organization and

[1] I have depended for this purpose on the surveys sponsored by the Indian Council of Social Science Research (1972–74; 1973), Sheth and Patel (1979) and Ganesh (1981).

influenced people's behaviour and performance at work. This concern led sociologists to devote their attention to the social and cultural characteristics of the people manning organizations (for instance, Lambert 1963; Subramaniam 1971; Niehoff 1959; Prabhu 1956). Some scholars were concerned with the behaviour and attitudes of people in relation to work and the work organization (Lambert 1963; Sharma 1974; Vaid 1968) and the interaction between work roles and social roles (Oommen 1978). A significant proportion of studies of behaviour and attitudes were related to industrial work, particularly with concrete problems of performance such as productivity, absenteeism, discipline and shift-work (Sreenivasan 1964; Sharma 1970; Shri Ram Centre for Industrial Relations 1970; Vaid 1967). Another stream of organizational studies covered important socio-psychological processes such as leadership, communication, decision-making and motivation (Sinha 1979; De Souza 1976; Bhat 1978; Basu and Patel 1972; Chaudhary 1978; Pestonjee and Basu 1972; Chowdhry 1970). These studies were usually governed by practical considerations, as they were based on the need to identify leadership styles, communication patterns and motivational strategies conducive to the performance-related objectives of the organization.

The various types of organizational studies mentioned in the foregoing have taken into account specific segments of social relations within organizations. A few studies, on the other hand, have sought to analyse whole organizations in a sociological perspective. The well-known Tavistock studies of the Calico Mills (Rice 1958, 1963) constitute the most systematic and comprehensive attempt in this direction. Rice and his colleagues examined the division of work and authority at various levels of the organization in the context of the technological as well as the human and social environment within which the organization existed. They introduced changes in the work organization and authority structures in response to the constraints generated by the socio-cultural bonds among workers and managers. Such reorganization, in the researchers' view, led to the optimization of the organization's effectiveness in terms of the primary task for which it was established. These studies, along with others in Britain and elsewhere, contributed to the sociological view of an organization as a system consisting of interactive social and technological factors. Sheth's study (1968) of an industrial organization analyzed the formal division of labour and the hierarchy of status and authority in the formal organization in the context of the structure of social relationships in the community outside the organization as well as in the context of the network of social bonds informally developed by workers and managers as a result of shared work experience. Similarly, Baviskar (1980) examined the social relationships within a cooperative enterprise in relation to the social and political divisions among workers and managers as well as among the political leaders and trade unionists associated with the enterprise.

Most studies of organizations, whether holistic or partial, share a common premise of sociological theory, namely, that an organization is a rational-legal system of tasks, authority and rules in the Weberian sense. The rationality of the system is usually conceived in terms of the goals of the organization as set by its founders, owners and managers. Those who are recruited into the organization for specific tasks are explicitly or implicitly expected to contribute to the managerial goals. Any behaviour among individuals or groups which is incompatible with organizational rationality is examined in terms of its dysfunctional consequences for the organization. At the same time, the satisfaction of human needs and aspirations based on psychological and cultural factors is assumed to be a major determinant of rational behaviour. Hence, sociological research in organizations is essentially geared to the task of identifying the various cultural and organizational forces impinging on human behaviour and performance at work. Such research is then used to pave the way to adapt management styles, communication patterns and reward systems to human social needs towards achievement of the organization's rational goals. The people manning an organization are thus regarded as integral parts of its rational system. This theoretical perspective provides the main foundation to the academic and managerial approach labelled 'human relations', which has become an important part of organizational thinking in modern industrial civilization in the West as well as in India.

The systemic approach to the study of organizations as summarized in the foregoing serves to highlight the integrative and collaborative aspects of organizations as on-going concerns in society. It stresses the function of collaboration among various sections of people involved in organizational activities for the effective performance of assigned tasks. Such an approach has considerable academic and pragmatic value in terms of its emphasis on the forces of consonance, consensus, harmony and integration within the social reality of organizations. However, this approach has resulted in an unwillingness among social scientists to take cognizance of another aspect of the same social reality: the forces of dissonance, dissensus and conflict among social groups, classes and categories. A few studies of organizations which focus on evidence of overt conflict deal with the most unambiguous and well-known form of conflict—the strike. However, the basic assumption about an organization as a harmonious, collaborative and integrative system leads the social scientist to explain conflict as a product of interpersonal and intergroup dynamics. This would include the assumptions made by managers and workers about each other, the patterns of communication between them, the feelings of alienation experienced by people under mass-production technology, and such other human and social factors (Dayal 1972; Dayal and Sharma 1970). Conflict is thus regarded as a manifestation of pathology in the social system of the organization. It is believed that such pathological disturbances can be removed by making suitable changes in the behaviour patterns, communication strategies and leadership styles of the various groups, especially the top management which is responsible for directing the system towards its rational goals. Ramaswamy's study (1977) of trade unions and workers in Coimbatore, and Mamkoottam's study (1982) of trade unionism in the Tata Iron and Steel Company, are perhaps the only serious attempts to deal with power and conflict within organizations with adequate objectivity.

The systemic approach leads to a partial and truncated view of organizations based on a specific conventional theoretical perspective underlying the discipline of sociology. Let us briefly review this perspective and examine the need and availability of an alternative perspective for an adequate sociological understanding of organizations.

SOCIOLOGICAL PERSPECTIVES

Sociology should be regarded as a scientific discipline concerned with human social behaviour. The term 'scientific discipline' is used here to stress that sociology, like other comparable disciplines of knowledge, implies a continuous search for consistencies and continuities in its chosen field of social reality, regardless of the probability of discovering general laws as defined in natural sciences. The main task of sociology is to analyze the various normative and factual aspects of social interrelations among people, and examine them in relation to one another with a view to explaining the social conditions under which specific forms of social behaviour and interactions occur. While such explanations would be primarily based on observations of concrete situations of social behaviour, the sociologist should look for explanations which can be generalized beyond specific events and situations.

The sociologist should choose an appropriate perspective to achieve this goal. Obviously, a significant part of social behaviour consists of patterned interactions reflected in the relatively durable parts of society, such as groups, norms, institutions and values. This has lent the facility to sociologists to look upon the social order as an integrated system with functionally interrelated parts. This integration theory (also known as functionalist theory) which, as Dahrendorf (1959: Ch. 5) states, has dominated sociological thinking, stresses the normative order in society, assuming a state of stable equilibrium among its components. Evidence of deviance from or challenge to the normative order is regarded as pathological, resulting in temporary states of disequilibrium. The dynamic and changing forces encountered in social reality are subordinated to the normative order.

This dominant perspective in sociology has, of course, contributed a great deal to the understanding of the normative aspect of society. However, it suffers from severe limitations, as several critiques (for instance, Lockwood 1956; Dahrendorf 1959; Rex 1961; Giddens 1968) have pointed out. The most serious drawback of the integration theory is that it neglects the factual order of society reflecting conflicts of interests and objectives among groups, classes and categories of people. These conflicts arise from the unequal distribution of scarce resources, including wealth, status and power. Another important argument against the functionalist perspective relates to its neglect of the historical forces underlying the contemporary normative order in any society. Moreover, the integrationist view has the effect of hypostatization of the existing normative order, and an implicit acceptance of its superiority over alternative models of social structures. Functionalist sociology thus tends to be partisan on the side of the existing social arrangement.

On the Indian scene, some scholars have recently questioned the integrationist-functionalist perspective. Those who have dealt with this subject share the view that Indian sociology is predominantly functionalist in its perspective, although a large number of sociologists have moved away from classical functionalism and combined it with historicism, and also paid some attention to power and conflict. The predominance of the functionalist approach is ascribed to the professional and intellectual dependence of Indian sociologists on their British and American counterparts, a dependence which is condemned with romantic anger by expressions such as 'thoughtways and workways of the colonial virus' (Dube 1977: 11), 'conceptual and methodological baggage of the Western social science' (Singh 1973: 15), and 'implanted by the colonial rulers as an administrative appendage' (Momin 1978: 160).

The awareness among Indian sociologists of the inadequacy of the functionalist approach has led some sociologists to plead for alternative approaches. A clear alternative perspective is developed and articulated by the Marxist sociologists. Desai (1981) has stated the value of the Marxist perspective vis-a-vis functionalism in a forthright manner.

> Indian society is subjected to a conscious transformation and change in a specific direction by policy makers. The social scientists pursue their researches of this changing social reality on the basis of accepting a historic, static, synchronic, structural-functional model based on an equilibrium system and stability models.... It is my submission that the paradigm evolved by Marx, if adopted consciously, even as a heuristic device, would provide...[an] alternative approach for conducting fruitful and relevant researches about Indian society. The Marxist approach adopting the criteria of taking property relations to define the nature of society will help Indian scholars to designate the type of society, the class character of the State and the specificity of the path of development with all the implications (Desai 1981: 8–13).

The message sought to be conveyed by the Marxist sociologists is clear: sociology should be regarded as a part of Marxist theory and philosophy. All social relations and group processes should be examined, analyzed and explained in the framework of property relations and ownership of the means of production. As the key concepts in functionalist sociology center around the assumption of the normative order in a state of equilibrium, the key concepts in Marxist sociology center around property and class relations. Both approaches contain an element of dogmatism and are therefore useless for scientific sociology as I have defined earlier.

Sociology, in my view,[2] needs to draw upon all aspects of social reality in crystallizing a perspective. It has to move towards generalized explanations of social behaviour. The normative order (including institutions,

[2] This view is based on the theoretical contributions made by scholars such as Lockwood (1956), Rex (1961), Giddens (1968) and Collins (1975).

values and culture) undoubtedly constitutes an important part of sociological studies as it provides valuable information about the social superstructure. However, the major concern of the sociologist should be to understand the relation between the normative order and the factual order, 'the whole of man's experience as a member of society in this world, here and now' (Beteille 1974: 100). This latter aspect of the field of sociological inquiry inevitably draws the sociologist into the various social, economic and political interests which divide and unite people into interest groups. One of the basic facts characterizing interest groups is the unequal distribution of status, wealth and power in society. This unequal distribution of material and non-material resources gives rise to relations of authority and power whereby groups with greater command over resources tend to actually or potentially coerce those with less command over the resources. The factual order of society underneath the normative order is thus characterized by relations of power and conflict which should constitute a major focus of sociological attention.

The balance of power is not fixed forever in modern society. Those who have more power usually strive to retain or enhance their power. Those who have less power, on the other hand, often try to grab power from those who have more. Human beings, individually and collectively, resist various degrees and different forms of coercion. Resistance may vary from passive submission through non-cooperation, bargaining, and open hostility to physical violence. The balance of power may therefore shift according to the dynamics of interaction among various interest groups.

Power and conflict are endemic in human society and should therefore form the central theme of sociological research. The normative order in many ways (through rules, rituals and superordinate agencies for conflict resolution, such as courts and arbitrators) provides the means for the resolution of conflicts of interests among groups and helps them to collaborate. However, these normative devices largely serve to contain or reduce conflicts; they cannot ever abolish conflict. As Lockwood states, 'The very existence of a normative order mirrors the continual potentiality of conflict' (1956: 137).

A PERSPECTIVE FOR THE SOCIOLOGY OF ORGANIZATIONS

Against the background of the preceding discussion on sociological perspectives, I shall briefly suggest an approach to studying organizational phenomena. Any sociological inquiry into an organization should of course include its normative aspects. It should take into account the stated goals, the division of work, the hierarchy of statuses and roles, the structure of authority and responsibility, the rules, procedures and conventions, the informal relations among people, and the technological, social, economic and political environment of the organization. One should study the processes of boundary maintenance, socialization, institutionalization and conflict resolution within the organization. One should examine how the various internal and external forces interact with and influence each other and influence the avowed goals of performance, efficiency, productivity, profitability, and so on. One should avoid the common tendency among researchers on organizations to analyze organizational behaviour essentially in terms of managerial rationality. The reality in organizations should be examined as experienced by the various sets of actors participating in it.

The experience of actors participating in organizational reality would vary, to a considerable extent, according to their relative positions in the power structure of the organization. It should therefore be recognized that a sociological analysis of interconnectedness among the various normative aspects is a necessary but not a sufficient step in sociological understanding. Adequate attention needs to be paid to the distribution of power within the organization, the implications of the power structure for the various categories, groups and classes of participants, and their behaviour in response to the distribution of power. The first major question regarding power in an organization is to define it contextually. The relative significance of power over fellow human beings and control over material resources, knowledge and information may vary among types of organization. The significance of supervisory control over, say, 100 subordinates for the supervisor and his performance is likely to be quite different in a political party, a university, a consumers' cooperative agency, a small family business and a large industrial complex. In sociological terms, power over the behaviour of other people who may react in many different ways constitutes a crucial aspect of an organization's overall control structure. The phenomenon of power therefore needs to be comprehended in relation to the people involved in it.

The second question regarding organizational power relates to the structural and cultural factors which act as its determinants. The basic determinant of power in a formal organization is the normative order governing the formal distribution of authority and responsibility. Another source of power is ownership of wealth, which often facilitates the individual or the group (for instance, a financial corporation or government) owning such wealth to dictate terms and actions to those who formally govern the organization. A third source of power consists of knowledge and experience relevant to the objectives, tasks and technology characterizing the organization. Technocrats and experts are known to wield considerable control over decision-makers at the highest levels in organizations. The fourth source of power is the possession of information about the crucial elements and situations in the organization. Thus, relatively junior officials handling financial or market information often enjoy power not only in relation to other groups within the organization but also in relation to outsiders such as clients and trade unions. The fifth determinant of power is the status occupied by a person in the social hierarchy outside the organization. For instance, those who belong to higher castes, more 'respectable' families, and prestigious educational institutions often secure obedience and conformity from colleagues and subordinates more easily than do others. The sixth important source of power is the ability to articulate the needs, aspirations, concerns and frustrations of significant groups within the organization and lead them towards challenge to official power. Workers' leaders holding positions in legitimized trade unions wield this power by virtue of the formal authority vested in them by law. However, this type of power is often derived more from a leader's ability to deal with critical problems than from the formal authority conferred upon him by law or agreement. In many cases, people wielding real power over workers and managers in the trade union context possess little or no formal authority.

The sociologist concerned with power in organizations thus needs to grapple with the various sources of power, as illustrated in the preceding paragraph. A detailed understanding of these forces is essential in the analysis of the distribution of power in the organization. The balance of power between the chief executive, the board of directors, and the various strata of managers, workers and union leaders in a business organization can be meaningfully comprehended only if the power derived from the formal, informal and environmental (cultural, political) sources is taken into account. Power understood and explored narrowly in relation to control over the performance of the main task is likely to provide only a partial understanding of the reality.

Another important issue regarding organizational power pertains to how it is used by those who possess it. Under what organizational, economic, political and social conditions do people in power subordinate their personal and sectional interests to the interests of others, including superiors, colleagues and the organization as a whole? Under what conditions do they pursue personal and sectional interests at the cost of the interests of others? For instance, it is widely believed that an important reason for the decay or virtual disintegration of many textile mills in India was the tendency on the part of the dominant entrepreneurs-cum-managers to manipulate organizational resources to serve their family interests to the exclusion of the interests of employees, consumers and other groups associated with the industry. Similarly, politicians, bureaucrats, trade unionists and other power wielders are believed to use their power to maintain and enhance it by manipulating, exploiting and coercing others into conformity. This type of behaviour is usually explained in terms of Michel's (1962) well-known iron law of oligarchy. However, as Ramaswamy (1977) has demonstrated with reference to a trade union, oligarchic tendencies among organizational leaders may be controlled by a membership with adequate political socialization and economic and social interest in the organization. We need to know much more than we do at present about the ways in which people in power use it in different conditions in the organization and in the larger society.

A question related to the use of power is the way power is made acceptable to those over whom it is used. In the Western cultural context, Etzioni (1961) has suggested a three-fold distinction of methods of reward by which conformity to power can be secured. Economic rewards induce the urge for maximization of economic gains and make commitment dependent on immediate pecuniary gain. Coercion induces alienation and results in the tendency to escape or attack the source of power. Normative rewards (such as appeal to larger social interests and ethical values) induce identification with the organization and hence moral commitment. These modes of reward and conformity constitute ideal types. The concrete behaviour of those in power and those who are controlled would combine these modes in varying proportions. Studies of the reward systems used by people in power and the conformity patterns among subordinates in the Indian context should therefore constitute an important part of the sociology of organizations.

The reward-and-conformity dimension of power in organizations merges into a larger sociological issue. Insofar as the distribution of power in an organization involves groups of people (like departments, categories of managers and workers, top management versus union leaders, and so on), how do groups at the two ends of power distribution interact? Under what conditions do the groups at the receiving end of power crystallize into interest groups and eventually into conflict groups? Under what conditions do conflict groups manifest loyalty to the organization in spite of coercion by groups in power? Under what conditions do they show indifference to assigned tasks? When do they manouvre for snatching power from others? When do they challenge power openly and resort to subversive methods to wrest power from others? These questions need to be answered in relation to specific organizational, social, economic and political situations. We should consider these issues not only with regard to power possessed by those in authority but also with regard to power possessed informally by certain work groups, status groups, groups of workers possessing special skills, younger generations of employees, informal cliques of strong men, informal cliques of union bosses, and such others.

All these issues in the sociology of organizations need first to be examined in specific organizational and cultural situations. The findings in diverse situations should then be compared to lead to generalized explanations pertaining to power and conflict in organizations in the Indian culture. This will also help in making predictions regarding the behaviour of various persons and groups in conflict situations in organizations. In this connection, it is interesting to note that Collins (1975: Ch. 6) has made a remarkable attempt to formulate a set of generalized sociological propositions on power and conflict in Western organizations. In doing so, he has interwoven the various strands of conclusions and generalizations from a wide variety of researches on the subject. Fortunately, in the Western context, 'If there is one area of sociology where serious cumulative development has taken place, it is in organizations' (Collins 1975: 286). Unfortunately, the performance of sociology on this score in the Indian context is depressingly poor. Generalized propositions regarding organizational power and conflict will therefore continue to be modelled on Western experience and theories until we acquire a sufficient fund of research based on Indian experience. Reassuringly, organizations do not constitute the only field of sociological or other knowledge which has to depend on Western experience and thought.

It is obviously not easy to study power and conflict in organizations. Researchers in this field are likely to be greeted initially by a 'not welcome' response from people involved in power relations, as discussions on power and conflict usually touch some of the most sensitive spots in their psychological and social existence. People who use power in relation to normative authority may not suffer from insecurity. However, those who enjoy power apart from or against formal authority usually feel insecure and threatened by inquiry relating to power and conflict. Also, studies of power and conflict relate to organizational phenomena which are incongruent with the current social values which eulogize cooperation, harmony, order and stability. Intellectually, any academic reference to conflict and power invokes the bogey of Marxism or anarchism, which is widely identified with disorder, revolution, subversion and such other activities popularly bracketed as anti-social. The sociologist interested in power and conflict therefore needs to devise appropriate methods and techniques to study such phenomena in specific situations. It should, however, be reiterated that a viable sociology of organizations can begin to develop only if power and conflict are treated as the focal points of sociological inquiry.

REFERENCES

Basu, K. S. and Patel, V. 1972. 'Method of Measuring Power Shifts', *Indian Journal of Industrial Relations*, 8, 21: 271–77.

Baviskar, B. S. 1980. *The Politics of Development*. Delhi: Oxford University Press.

Beteille, A. 1974. *Six Essays in Comparative Sociology*. Delhi: Oxford University Press.

Bhat, V. V. 1978. 'Decision Making in the Public Sector: A Case Study of Swaraj Tractor', *Economic and Political Weekly*, 13, 21: M.30–M.45.

Chaudhary, A. S. 1978. 'Downward Communication in Industrial Hierarchy in Public Sector Organization', *Integrated Management*, 13, 5.

Chowdhry, Kamala. 1970. *Change in Organizations*. Bombay: Lalvani.

Collins, R. 1975. *Conflict Sociology: Toward an Explanatory Science*. New York: Academic Press.

Dahrendorf, R. 1959. *Class and Class Conflict in Industrial Society*. Stanford: Stanford University Press.

Dayal, I. 1972. *Anatomy of a Strike*. Bombay: Somaiya.

Dayal, I. and Sharma, B. R. 1970. *Strike of Supervisory Staff*. Bombay: Progressive.

Desai, A. R. 1981. 'Relevance of the Marxist Approach to the Study of Indian Society', *Sociological Bulletin*, 30, 1: 1–20.

De Souza, A. 1976. 'Some Social and Economic Determinants of Leadership in India', *Social Action*, 26, 4: 329–50.

DUBE, S. C. 1977. 'Indian Sociology at the Turning Point', *Sociological Bulletin*, 26, 1: 1–13.

ETZIONI, A. 1961. *A Comparative Analysis of Complex Organizations.* New York: Free Press.

GANESH, S. R. 1981. 'Research in Organization Behaviour in India, 1970–1979: A Critique', unpublished Working Paper 372. Ahmedabad: Indian Institute of Management.

GIDDENS, A. 1968. '"Power" in the Recent Writings of Talcott Parsons', *Sociology*, 2: 257–70.

INDIAN COUNCIL OF SOCIAL SCIENCE RESEARCH. 1972–74. *A Survey of Research in Sociology and Social Anthropology*, 3 Vols. Bombay: Popular Prakashan.

——. 1973. *A Survey of Research in Management*, Vol. I. Delhi: Vikas.

LAMBERT, R. D. 1963. *Workers, Factories and Social Change in India.* Princeton: Princeton University Press.

LOCKWOOD, D. 1956. 'Some Remarks on the Social System', *British Journal of Sociology*, 7: 134–43.

MAMKOOTTAM, K. 1982. *Trade Unionism: Myth and Reality: Unionism in the Tata Iron and Steel Company.* Delhi: Oxford University Press.

MICHELS, R. 1962. *Political Parties.* New York: Free Press.

MOMIN, A. R. 1978. 'Indian Sociology: Search for Authentic Identity', *Sociological Bulletin*, 27, 2: 154–72.

MUKHERJEE, R. K. 1973. 'Indian Sociology: Historical Development and Present Problems', *Sociological Bulletin*, 22, 1: 29–58.

NIEHOFF, A. 1959. *Factory Workers in India.* Milwaukee: Milwaukee Public Museum.

OOMMEN, T. K. 1978. *Doctors and Nurses.* Delhi: Macmillan.

PESTONJEE, D. M. and BASU, G. 1972. 'Study of Job Motivation of Indian Executives', *Indian Journal of Industrial Relations*, 8, 1: 3–16.

PRABHU, P. N. 1956. 'A Study on the Social Effects of Urbanization', in *The Social Implications of Industrialization and Urbanization.* Calcutta: Unesco Research Centre on the Social Implications of Industrialization in Southern Asia.

RAMASWAMY, E. A. 1977. *The Worker and his Union.* Delhi: Allied.

REX, J. 1961. *Key Problems in Sociological Theory.* London: Routledge and Kegan Paul.

RICE, A. K. 1958. *Productivity and Social Organization: The Ahmedabad Experiment.* London: Tavistock.

——. 1963. *The Enterprise and its Environment.* London: Tavistock.

SHARMA, B. R. 1970. 'Absenteeism: A Search for Correlates', *Indian Journal of Industrial Relations*, 5, 3: 267–90.

——. 1974. *The Indian Industrial Worker.* Delhi: Vikas.

SHETH, N. R. 1968. *The Social Framework of an Indian Factory.* Manchester: Manchester University Press.

SHETH, N. R. and PATEL, P. J. 1979. *Industrial Sociology in India.* Jaipur: Rawat.

SHRI RAM CENTRE FOR INDUSTRIAL RELATIONS. 1970. *Human Problems of Shift Work: A Study of the Socio-Psychological Processes in Worker Adjustment to Shift Work in Industry.* Delhi: Shri Ram Centre for Industrial Relations.

SINGH, Y. 1973. 'The Role of Social Sciences in India: A Sociology of Knowledge', *Sociological Bulletin*, 22, 1: 14–28.

SINHA, J. B. P. 1979. 'The Nurturant Task Leader', *ASCI Journal of Management*, 8, 2: 109–19.

SREENIVASAN, K. 1964. 'Workers' Productivity and Outside Plant Life', *Productivity*, 12, 1: 27–34.

SRINIVAS, M. N. 1966. *Social Change in Modern India.* Bombay: Allied.

SUBRAMANIAM, V. 1971. *The Managerial Class of India.* Delhi: All India Management Association.

VAID, K. N. 1967. *Papers on Absenteeism.* Bombay: Asia.

——. 1968. *The New Worker.* Bombay: Asia.

5

The Dogmatic Foundation of University Life

F. G. BAILEY

INTRODUCTION

M. N. Srinivas, in a somewhat acerbic comment on Weber's views of Hinduism, dismissed the notion of a 'dogmatic foundation' (Singer 1973: 281). To me one of the most agreeable features of Srinivas' writing is his ever-present sense of empirical reality and an unwillingness to rise to levels of abstraction where logic rules alone and evidence is of no concern. Diversity, in other words, is a phenomenon to be honestly encountered and not to be swept away in a taxonomic game that is good for exciting the intellect but not for understanding what people do, or even the way they think.

In this paper, I will look for the 'dogmatic foundation' of university life insofar as this exists in the goals of the institution.[1] It is my pleasure to offer it in dedication to Srinivas.

[1] The material is drawn from interviews and informal conversations with academics and administrators in universities in Fiji, New Zealand, Australia, India, Britain, the

GOALS

Universities, being formal organizations and artificially created, are supposed to have goals. 'Formal organizations are explicitly instituted to achieve given objectives' (Blau 1964: 329). That is what distinguishes them from that other type of human aggregates, vaguely called 'organic' or 'natural', like the family or the community, which have no goal other than their own existence. On inspection families, communities, tribes and other 'natural' aggregates will be found to have functions—socializing children, economic activity, political cooperation and so forth—but these are not set out as goals for which the family or community was created. Indeed, it works the other way round—the activities associated with the various functions serve the community, but organizations are created in order to provide the activity. In short, one cannot conceive of an organization that does not have a goal or set of goals.

The word 'goal' suggests a direction in which one should be moving, and a statement of goals comes out as a list of directives as to what policy should be adopted and what action taken. 'Direction', indeed, seems to be a compelling metaphor.

> I promised...that I would attempt...to formulate...specific proposals, which...would in effect chart the course which we shall follow over the next three years or so. ... In this connection I am reminded that it is only those who do not know where they are going who can afford to proceed without a map of some kind for if you do not know where you are going—almost any road will get you there![2]

The metaphor of travel also prompts the reflection that it is impossible to move in several directions simultaneously. A set of directives must be internally consistent. Rules which tell you at the same time to

---

United States and Canada. It has occasional reference also to universities elsewhere, in places where my collocutors had previously worked.

This work was substantially aided by the award of a Guggenheim Fellowship. I am grateful to my colleague D. F. Tuzin for his comments, and to Marian Payne for preparing the manuscript.

[2] All quotations are edited transcripts from either taped interviews or documents. I have not, even in the few cases of published documents, identified the speaker or the place, considering their privacy more important than the provision of context. This is a necessity, albeit an unfortunate one.

march east and to march west cease to be directives, because you cannot be guided by one rule without disregarding the other. Consequently, it is felt, there must be consensus about goals.

> Colleagues and friends, I am convinced that there is a consensus on our general objectives—on where we wish to go.

But the consensus must be worked out through settling various lesser questions:

> Inevitably, there are many routes which could be taken, different forms of transport that can be used, and we even have a choice of travelling companions. In trying to plan our journey, the speed at which we wish to proceed is also a major factor. These matters can be discussed and hopefully will be seriously examined and assessed.

In that process the term 'consensus' undergoes a subtle change. It is no longer an unforced agreement—existing *ab initio*, so to speak—about goals: the new consensus requires a formal abandonment of any erstwhile goal not congruent with it.

> But there comes a time, even in academe, when decisions have to be taken. Once this is done, it behoves all concerned—all the voyagers—to cooperate fully.

There is a mildly minatory note about this. Consensus about goals, if not present from the beginning, must be created in its second form and the price of not doing so is to perish.

> While I do not wish to erect a bureaucratic framework or to limit the freedom enjoyed by any of us, I cannot, because of the responsibility trusted to me permit an untidy system to gain root. ... Many of you are not yet aware of the serious damage which some of our attitudes and utterances have done...and are continuing to do. Some of our benefactors have got to the stage where they will take no more buffeting, and I cannot in all honesty say I disagree with them.

So, while 'convinced that there is a consensus on our general objectives', the speaker evidently also felt that not everyone was sufficiently toeing the line, and he reminds his audience that the price of 'an untidy system' could be the loss of benefaction. There is an element of paradox in this demand, for the new consensus, despite its origins in prudent self-interest, should not be mere compliance, bought with the wages that go with the job. It should have a moral quality about it, a positive acceptance of the goals as good in themselves.

Some of this is reasonable and practical enough. If there is no consensus about goals, the various purposes of different individuals and sections within the institution being not consistent with one another, then these differences must be resolved so that decisions can be made and action can be taken, and the benefactors kept happy. But the process of working towards consistency and consensus requires that strange act of conversion by which at least some people must abandon a goal which had their first priority in favour of something else, which they are expected to endow with an intrinsic value which formerly it did not have.

My purpose is to examine this striving for consistency and consensus. Through examples I will show what dissensus about goals in universities looks like. I will ask what means exist for transforming the dissensus into consensus, and whether any of these means in fact encompasses the seemingly magical act of value conversion outlined in the foregoing.

'WHAT I WOULD LIKE TO SEE'

When you ask a university teacher or an administrator for what purpose the institution exists, the answers are varied and not consistent with one another. First—possibly because it is easy to describe and gives the speaker time to think—may come the official answer, the one which is printed in university documents, and which receives scriptural commentary each year in the formal addresses of Chancellors and Vice-Chancellors.

> The University was created as a research university, being established in...by an Act of Parliament 'to encourage and provide facilities for postgraduate research and study both generally and in relation to subjects of national importance'.

Next might come either the speaker's own opinions about what the institution should be doing, or a description of what (according to him) other people would like. Often the two statements or sometimes all three were placed in a dialectical form. The 'official' description provides the purchase required to push oneself into a different position. If this position is one occupied by other people, that in turn may provide the contrast required to make clear the speaker's own opinion. In the following two-stage answer, after stating that the institution was established to serve the region in vocational training and applied research, the speaker went on to say:

> Social science, and to some extent physics, are out of tune with the rest of the institution. They are interested in research and direct their attention less to vocational activities and the institution itself, than towards what is going on in the worldwide field of their own discipline. Such people find it strange that their attitude towards scholarship is not found in all parts of the institution.

Others made their own preferences clear by contrasting them with what was being done.

> What is the university doing? Our main function here is to manufacture a privileged elite. We do that very well. we divorce them from the public very well. How you design any university so as not to divorce them is a problem. I know you find taxi drivers in the US who are university graduates, but this region cannot afford to do it that way. The Extension thing overcomes it. The student talks to other people: he's doing a job and is in touch with reality while he is studying. That is the way we can learn to reach out to these people.

The disparity between the 'official' answer, the various other answers, and the description of what the institution really accomplishes varied between speakers and between institutions. But almost always the three- or four-step exposition was used to emphasize contrasts rather than similarities. Not unexpectedly, the failure to attain goals seemed to call more urgently for analysis than successes.

At that stage the conversation might turn to explaining why the goals are not attained and why one has no choice but to settle for a less than perfect institutional world. The explanations (to be considered later) invoke a variety of elements, which the speaker defines as external and intrusive by the criterion of the goals which he himself

sets for the institution. Thus the 'official' definition of the institution's purpose may itself be presented as one of the causes of the institution's imperfections. The actuality then is portrayed as a compromise between what that person thinks desirable, what others think is desirable, and what the economic, administrative, political and socio-cultural environment of the institution can or will allow to be done. Out of this emerges three kinds of statements about ideals: 'What I would like to see'; 'What in the circumstances I can reasonably aim at'; and 'What in fact is done'. I begin by describing the various forms which these several levels between 'perfection' and 'reality' can take.

Almost without exception, the purpose of the institution was said to be service to some higher entity than the institution itself. There was one dissenter who, after a mildly derisive reference to ideals, plunged straight to the third level:

> Tertiary institutions are organized anarchies. They do not have goals; or they do not know what they are; or they have a lot of goals at different levels. Drop a hat and the Principal will give you a speech about goals, but like any other institution, the institutional goal is first to expand, and if it cannot do that at least to stay in business.

That was the exception. All the others spoke of service. There were two kinds of service, corresponding to Gouldner's (1957) 'cosmopolitan' and 'local'. The former are those who see their task to be the advancement of knowledge and they measure their success by the opinion of others in the same discipline. A few are insulated even from the opinion of their colleagues and they see truth as an absolute and objective thing, the outcome of their own personal intellectual contest with formlessness.

The feelings of those whose eyes are directed towards 'the world-wide field of their discipline' is plain in the following examples.

> Here the major part of the expenditure is on research and research training, but not for people at the workbench level. Here they look to higher levels, the carrying forward of the grand plan. The early people who set up the university had that in mind: we were deficient in such people, those who had the talent having gone overseas.

If all were to be for the best, in a best of possible worlds, then for some people pure research would be the supreme activity.

> Without research there is no point to academic life; research is its whole meaning and is absolutely significant.
>
> My prime concerns are to affirm that creative scholarly and/or scientific endeavour is an essential part of an academic's calling, and to warn that if it is squeezed out too much, or neglected, it will be university undergraduate teaching itself which in ten years' time will be the chief victim.
>
> This place is formally and officially defined as a research outfit. It is not obliged to take graduate students. If any faculty member teaches undergraduates, that is because he likes to do so. They may—and often do—decide to cut scholarships in the interest of faculty research or Fellowship appointments. Research money is built into the institution's grant and our people cannot apply to other national funds. They can do consulting type of research, but few in fact do. Most of them are not equipped for applied work of that kind, and that was not the intention of those who founded the institution.

One speaker put the point defiantly, surely with tongue-in-cheek.

> Land grant colleges are sometimes kept closely responsive to the legislatures which set them up, and this is likely to produce demands for teaching and service in various forms. Other places, like the one where I work, consider themselves responsible to no one except themselves and they favour only those kinds of research which are of absolutely no use.

Opposed both to the self-contained scholars and others who require the support and approval of their fellows are those who regard knowledge not as an end in itself, but as a benefit to humanity—a tool in the service of other people.

> There seems to have been some confusion in the minds of people here about what universities ought to be doing. One thing is the pressure to provide vocational courses—things that are useful. The last place where I worked was set up as a university of technology and it was heavily involved in areas of direct use, particularly in the applied sciences.

But it is hard to find agreement about where the line between pointless scholarship and useful research should be drawn.

> I clash with him a great deal: he values vocational and practical things. Science is right enough: anything to do with the arts or humanities he thinks of as frivolous.

The following speaker, a social scientist, made his position very clear.

> After those years in a developing country I cannot stand research projects which are aimed solely at developing theory. They make me very uneasy. Theory should arise out of a research project which is primarily about the real world. This question really affects me. The notion that one goes to some part of the world to look at asymmetrical alliance I find really immoral. Let us have something that makes sense to the country where it is done. The proper thing is to use theory to understand the real world rather than using the real world to play theoretical games. I was trained to play those games, but having spent those years in a developing country, I had a change of attitude.

There are more immediate forms of service to others. The institution ministers not to the advancement of knowledge but to the needs of one or another kind of community.

> This place was not intended to be just another institution for tertiary education, but rather one with a special purpose—to serve the local community, training people, and doing projects that would serve local business and industry. There are distinct expectations from the local community and its leaders.

The service which the universities provide is sometimes depicted as symbolic rather than practical. Here is an American speaking of universities in Australia where he worked.

> Universities here were not founded like the land grant colleges to serve, but rather as an ornament to society. You have to have a university, like you have to have a GPO, a cathedral and a railway station. The professors were remote and elite people. In the 1930s there were only 25 of them in this university, helped out by lecturers and tutors. A young man would come out to a Chair from England—or better from Scotland—and would be received into the best society and would marry into one of the wealthy families.

This statement has a mythological flavour, but the notion that a university is a cultural embellishment is not uncommon, and most academics would find comfort in the idea that local people respect and admire the institution. But they would certainly feel demeaned if that were presented as the main reason for their existence. At least let them rather be a source of enlightenment than a mere ornament. Still less would they like the idea that they and the college were pawns in the politicians' chess game.

> I know a little place where the lawyers started agitating for a local college. They put up a temporary building and hired unqualified and incompetent teachers and, in the end, they pressed and lobbied long enough until the politicians gave way and it was recognized as a college and got its share of public money.

Another form of service to the community is consultancy. The university is a repository of expertise, upon which local leaders and politicians can draw. This is not the same as applied research in which the expert defines the problem: in consultancy the problems are defined by others and the expert is a technician who builds to another's design.

> We should do much more consultancy. I have got several things going now, and much more consultancy gets done here than at the average university. Vastly more informal consultancy. A range from being pulled into a corner at a cocktail party by senior government officials to being on committees.
>
> The government has an absolute hold financially. For a long time they never believed the place had much value. Why don't they just keep quiet and produce teachers? Recently they have come to look on us more as a developmental arm. It is in fact a major bloody resource. This is by far the biggest pool of so-called talent in the region.

Universities also teach students, having, unlike a research institute, an obligation to train young people. This requirement is not in dispute, but there are widely differing opinions about the nature of the training which should be offered.

At one extreme is an attitude which belongs with purer-is-better research.

> He professed one of the remoter languages of western Asia. He was one of two or three experts in the world, and much respected by linguists. When the academic year began, he would go into hiding for two or three weeks in case someone turned up wanting to study that language. He wanted only one pupil, he said: a first-class scholar who could eventually take over his position.

This story does not have a ring of truth in it, in a literal sense. But it does accurately portray an attitude towards education: the business of the scholar, when involved in education, is to train other scholars like himself.

But that may not be what the tax-payers intend: they want their sons and daughters to be trained to make a good living for themselves and to be of use to the community.

> The intention was that this place should respond to community needs for technological education by providing vocational training. It grew out of and drew staff from the local technical college, and it absorbed a school of engineering and various other institutions which dealt with medical technology, public health and so on. It was generously funded by the state and later it got federal funds.

This feeling that one must be in touch with the community goes beyond teaching, in some cases, to active participation in community affairs. The teacher, it is felt, has an obligation to make available his experience and his intelligence.

> There are relevant things that I feel I must do for the community. I am involved in many practical things to do with social questions. Just this month I am arranging a conference, for example, for the Law Reform Society.
>
> I had close connections with a Director of Education for this region and I am responsible for [my discipline] being taught in the colleges of technology and in various teachers' colleges. I believe that teachers should have some knowledge of their own society, particularly of its racial features. So we have in fact been very receptive to the demands of the society in which we live, but we have taken care never to be subverted by those demands. We have been flexible. Indeed we create many of those demands ourselves, as in the part we play in training teachers.

This benign and paternalistic outlook is upheld by some people and mildly disparaged by others, but it is not a matter of great controversy. Whether one should educate an elite or whether one should—or even could—make tertiary education widely available, however, is a question which arouses much stronger feelings. One example, given earlier, was the man who favoured extension teaching. Here are two others.

> The real problem seems to be the expectations of the community at large, and of the student and the government. First of all it is a highly elitist place. Students and the public at large think that if you get into the university you *ipso facto* deserve a good job. This is carried to extremes by the government, who maintains salary differentials throughout the working life of an employee according to the class of degree that they got. It can make as much as a 20 per cent difference to your salary, and this goes on through life.

Even within the university, the wheat may be separated from the chaff.

> Government has gone right into not just the organization of the university but also the structure of the courses. They have made it much harder for students to specialize, apart from the brightest 10 per cent who get into the Honours course where they concentrate on one subject. My interpretation of this is that the government does not want very well-educated trained thinkers at large in the population. It wants no more of those than can be absorbed into the Civil Service, which is just about where all the Honours graduates go.

The business of the university, in other words, should not be to turn out specialists but to train young people to think for themselves and to question established knowledge and established procedures. To drill them in known routines is not sufficient.

> I imagine that is true for any third world country—the use of the university as a kind of culling mechanism for the society at large. The government misses the point of the university when it tries to use it simply to produce specialists. That is what polytechnics are for.

We are about to ask how these varied goals relate to one another. But, before that, two questions must be answered. First, how do these divergences come about in a single institution? The principal cause is the academic habit of out-breeding (having your pupils go elsewhere to teach) and the ancient academic institution of the wandering scholar. We can call it the *alma mater* factor, and it comes about in the following way. From one's place of training (and from one's generation) is derived a set of assumptions about what a university should be. This model to some extent survives migration, becomes somewhat divorced from context and remains a partial guide in new places where the ambience may be less suited to it. Add to this varying socio-political philosophies (some of them, no doubt, a legacy of early academic training) and the outcome is a firm conviction that the ivory tower or the training ground for young revolutionaries or whatever else is the only proper academic goal.

Second, why do the discrepancies matter to any particular institution? They matter only because the people concerned have strong opinions about them, not only about particular options, but also about the existence of contradiction itself. For some it is unambiguously an institutional sickness that must be eliminated; for others it is a sign of health and vigour; only to a very few is it a matter of total indifference. Therefore, if one's aim is to understand how such institutions work, the disaccord over goals must be taken into account.

I shall now examine the ways in which this disaccord might be handled.

DISACCORD: REAL OR APPARENT?

One way to handle disaccord amounts in essence to proving that the diversity is only apparent and is encompassed, Dumont-style, within an overarching unity.

The framework used in the preceding section to list the various institutional goals encourages a Manichean outlook: one side is good and the other bad, and they are in opposition. There is a choice between good and evil and nothing in between. The scheme is summarized in Figure 5.1

Is there concealed in this apparent diversity a set of central values which might 'govern' universities? The point of the question, as noted earlier, is that orderly government is less likely to emerge from directives which point simultaneously in different directions.

One way to demonstrate that there is a single internally consistent set of values might be to show that all the terminal positions (A2, C1,

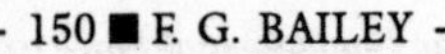

Figure 5.1
Institutional Goals

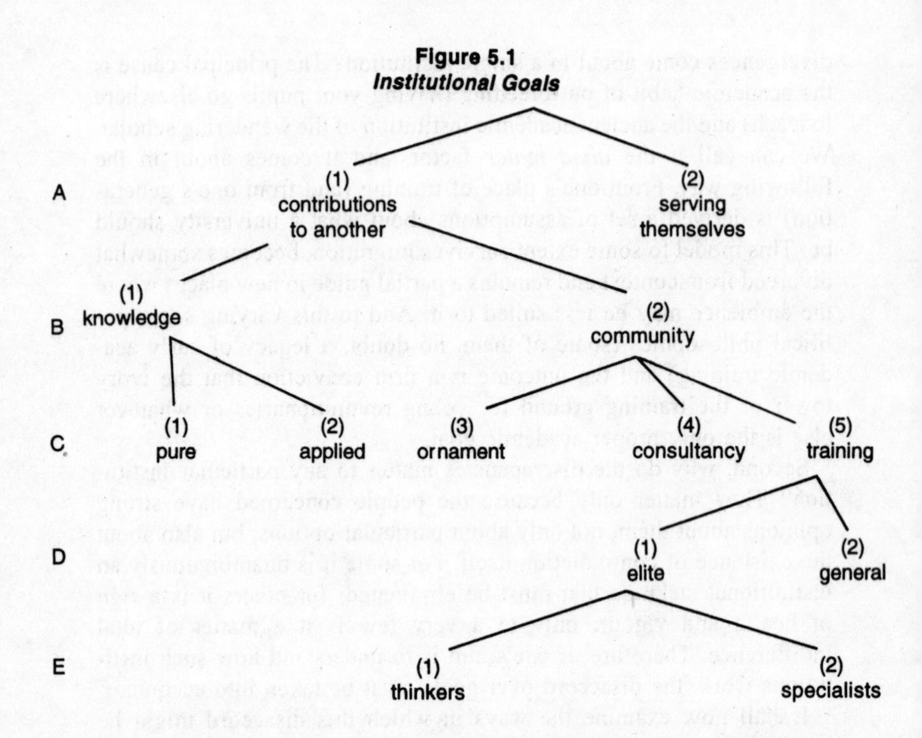

C2, C3, C4, D2, E1 and E2) are transformations of a single item. On the ground we have oak and ash and beech and birch, but in the word 'tree' they all become one. Of course one can do this as an analytical exercise for virtually any aggregate of items, like ships and shoes and sealing-wax and duodenal ulcers: but the exercise will be trivial if it has no substantive implications. Thus, in the present case, the form of the figure suggests that all these eight items are varieties of what is at the top. But the top is the word 'goal', and this has no substance to it, for what matters is not that institutions have goals (they do so by definition) but rather what is the nature of these goals. As soon as substance is inserted discrepancies and inconsistencies reappear.

The result is no better if one inserts the word 'service', which apparently does have some substance to it (one can imagine institutions which proclaim goals other than service). Wherever one goes in Figure 5.1 (even A2), something is being served. But once again we are landed in a triviality. The assertion is of very little analytic interest (although in some situations it might do useful rhetorical work) until one has added more substance by asking: service for what? The disaccord then reappears.

If, then, one despairs of putting anything of practical and substantive significance at the top, perhaps the diversity and the potential disaccord might be reduced to something less than the eight terminal positions in Figure 5.1. Inasmuch as we can group them and make fewer categories of them, are we then nearer to finding a set of internally consistent central values? Figure 5.1 does provide such a reduction: in place of the eight items are the three kinds of service: to the institution (A2), to the discipline (B1), and to the community (B2). The diversity is indeed reduced, but by that very process the intensity of disaccord is likely to be increased. In general the higher one pushes the debate in a preference tree of this kind, the more one signals intransigence and an unwillingness to give in order to get. The higher one rises the more are the values that have to be compromised, and the more numerous are the people who have to be 'squared'. To borrow a Shils metaphor (1975: 168), because it is fittingly suggestive, there are no longer several low peaks of disaccord but three large ones, and it is harder to climb down from a high peak than from a low one. In short, the exercise of analysis does not, in any practical way, bring unity to the set of central values: it reduces the diversity but intensifies the oppositions.

A second difficulty in this kind of analytical approach is that there is more than one way to make the reduction. For example, given that many academics value independence and insist on their right to pursue truth wherever it leads them, perhaps that might be used as a criterion for analyzing the eight items into fewer categories. The outcome is not the same as when the criterion of who should benefit from service is used. If one asks whether the academic should set his own goals and be accountable to 'truth' (represented by himself and his fellow scholars) or whether he must answer to outsiders, various adhesions or affinities appear across the different categories. The independent scholarship values in C1 is in practice repeated in C3, is not entirely absent in C2 (insofar as the research worker defines the problem himself) and emerges as the defining element in position E1, in which the function of a university is to teach the young to think for themselves. Similarly, as we saw earlier, *one* version of C5 (the professor who wanted only one pupil during his career) is an adjunct of position C1, serving to maintain C1 as the *status quo*.

When one has two or more versions of the componential structure of a domain of this kind, there are various and much disputed ways for arriving at where 'psychological truth' must lie. My argument, to be made more fully later, will be that truth is irrelevant in this case—what people *think* is of no significance in the enquiry, one pays attention to what people *say*. The immediate relevance of discrepant analyses is that they diminish the chances of finding unity in the set of substantive central values by increasing the dimensions of disagreement and therefore also the possibilities for disaccord.

At least at this level—what universities should be doing—it seems unlikely that this type of analysis will provide a unified set of central values.

But, it might be argued, the test is not fair. The quotations come from different institutions, which are placed in varying situations, and one should not therefore expect to find a consensus. It remains possible that one may find such accord within the bounds of a single institution. The logical exercise of trying to find the single *fons et origo* that will eliminate diversity is indeed pointless, but empirically one will find a broad agreement about what should be done within the walls of a particular college or university. Despite the diverse origins of individual scholars (the *alma mater* factor), one can reasonably expect a general consensus on central values, both in the sense that they are espoused by the ruling authorities and are generally accepted by ordinary people in the institution.

The first part of this claim is true. So long as one confines the notion of 'centre' to the 'ruling authorities' (Shils 1975: 4) together with the charter, the statutes and so forth, there very often is such a consensus, certainly in what gets said, if not always in what gets done. If one means by 'central values' a set of goals and directives about which there is general agreement among the administrators and the academic faculty, irrespective of age or scholarly interest or previous experience, certainly there are some institutions in which, for a time, there is a high level of consensus about goals. The authorities take care to recruit only suitable people. But there are few institutions which can continue to display such uniformity. Mistakes in recruiting are made: fashions change and those taken on ten years ago (now with security of employment) would not be recruited today, for the 'consensus' has changed. Most universities have something of this patchwork-and-overlap quality about them. Then the consensus disappears and Figure 5.1, with at least some of its eight mutually inconsistent terminal positions, would cover the range of disaccord in the institution.

But, it might be argued, this is true only at the first level—the 'utopian' one of 'what I would like to see'. As soon as one comes down to the second level, that of the practical, certain features become apparent, all of them tending to reduce the level of incompatibility. First, it becomes clear that the effort to reduce differences by the method of logic is not only without practical benefit (because it cannot be done except at the price of ascending to formal trivialities), but it is also inept because most of the differences in Figure 5.1 arise not from logical contradictions but from practical incompatibilities. Research and teaching are not in any logical relationship to one another, certainly not one of contradiction. Rather they are incompatible in that, given limited resources, both goals cannot be achieved. They are also incompatible because different people in the same institution want different goals to be pursued. One has to make a choice and establish preferences. So we are dealing not with pure reason but with practical reason; with stating preferences, ranking goals, and defending one's preferred rank order against other people who want different choices made; in short, with interests.

THE MECHANICAL ELIMINATION OF DISACCORD

It might be argued that if this is a case for practical reasoning and establishing preferences, then there are well-established methods for doing so, and these methods should serve to identify the central guiding values. One has to do no more than poll those concerned, and the theory of elections provides an amplitude of methods for conducting such a poll. I have not tried the experiment, not only for lack of resources during the period of research but also because I think that not even the best result would in fact reveal those central values which are supposed to govern the institution. On the contrary, it seems that the very possibility of conducting such a poll rests not on an ultimate numerically-ascertained level of consensus about what the university should be doing, but rather on the existence of another kind of central value, one which governs not academic goals but rather modes of political behaviour within the institution. This value is collegiality.

The notion that the members of a university might in a once-for-all, comprehensive fashion poll themselves to find a consensus of values is somewhat fanciful. But it is done in a piecemeal and fragmentary fashion every day and every time a committee meets, and occasionally disagreements on particular issues are in fact made the subject of plebiscites. There is, therefore, a practical use in thinking about the effectiveness of such procedures in creating harmony and eliminating disaccord.

When surveys of this kind are conducted, although they may set out to find 'the truth', in the end they produce only ammunition for debate, being yet another way of trying to persuade the unconvinced: 'If 90 per cent want it, how can it be bad?' to which one may respond either that it can, despite the 90 per cent, or that the figures are wrong.

One reason why polling will not create harmony out of a discord of values is that the methods of polling are themselves matters of dispute. Following the essay by the Rev. C. L. Dodgson (Black 1963: 216ff), these are some of the methods at one's disposal: simple majority; absolute majority; elimination (candidates voted on two at a time, the loser being dropped each time from the competition); elimination (all candidates voted on, the one with the fewest votes in each round being dropped); and the 'method of marks' (each voter ranks the candidates, gives 0 to the lowest, 1 to the next lowest, 2 to the one above, and so on: the candidate with the most marks wins); and there are other methods.

How does one select a method of polling? Dodgson shows that all these methods (except the last) can elect a candidate whom the majority of voters did not want chosen. For example, suppose four electors (I–IV) choose between (A) pure research, (B) elite teaching, (C) mass teaching, and (D) applied research, and suppose the following is the pattern of their preferences:

I	*II*	*III*	*IV*
A	A	B	D
B	B	C	B
D	C	D	C
C	D	A	A

then, by taking an absolute (or simple) majority as the method, A would be chosen despite the fact that it is the lowest preference of half the electors, whereas B, which never falls below second place, should, given the intention of getting as near as possible to a consensus, have been the winner. Similar peculiarities eventuate in all the other methods of selection, except Dodgson's favoured 'method of marks'.

Conducting a survey is a technical matter and the investigator is free, one hopes, to select the method that will bring him nearest to the 'truth'—that is, the actual distribution of preferences in the institution. If he is convinced by Dodgson, then he will use the final method, for the outcome will produce that goal which has most approval and least disapproval, and which might be termed, in a rather loose way, a kind of consensus.

But what kind of consensus is this? The method of marks was described by Borda in the second half of the eighteenth century. Someone pointed out that the astute voter, having placed his favourite candidate at the head of his schedule of preferences, would then put the strongest opponent at the bottom. 'This would give great advantage to candidates of mediocre merit, for while getting few top places they would also get few lowest places. Borda, who had foreseen this, said, "My scheme is only intended for honest men"' (Black 1963: 182). A similar device intended to outflank the system is to bracket all candidates together except the favourite, leaving him with the score of 1 and the rest with 0. Dodgson amended the rules to defeat this manoeuvre by giving all the bracketed candidates the mark that would have gone to the highest if they had not been bracketed, thus encouraging the voter to draw up a 'genuine' schedule of preferences.

This device does not remove the objection that the system favours mediocrity. The less important a value the more likely it is to find agreement. When we are thinking of goals or policies or courses of action, it means that in practice the ardour that might have gone into implementing goals linked to 'irreducible values' (which have by this method in fact been reduced) is diminished. This is particularly so if the gap between the favoured goal and those below it is a very large one. Galton (quoted in Black 1963: 88) dismisses averaging in favour of the 'middlemost', because averaging 'would give voting power to the "cranks" in proportion to their crankiness'. But there are occasions when cranks and fanatics or strong-minded people confident in the rightness of their position have their uses.

This should alert us to the fact that while a survey is a technical matter, conducting a poll with a view to taking action is quite a different affair. It is political, not purely technical, and technical considerations tend to be subordinated to political ends, as in the instances of bracketing all but the favourite or putting the most dangerous opponent at the bottom of the schedule. To begin with, the method of polling is not necessarily selected only on the basis that it is sensible, that it will come nearest to whatever is the 'truth'. Those concerned have a stake in the outcome. Many of them already know where 'truth' lies—or at least where it should lie—and they are inclined to work out which method of polling is most likely to produce *their* rank order as the 'true' one and to scheme for its adoption. Already at this stage—choosing a method of polling—objectivity has flown out of the window and the clash of opinions, instead of waiting passively to be silenced by the poll, is banging it out of shape. In short, a poll begins by producing another dimension of disaccord before it can be turned to resolving the discord of goals.

But let us suppose that they get over this hurdle and decide to use the method of marks. Will its numerical manipulation lead us to consensus?

There are reasons why, in practice, it would not. The goals I am assuming purport to be fundamental and irreducible. If one selects a candidate for a fellowship (the occasion for Dodgson's essays), the value set on him is not of this irreducible order: one generally has a reason for choosing him above the others—'He got a better degree', 'He went to a better school', 'We don't want another one of them in the Common Room'. But the preference for pure research or to devoting one's life to training an elite are ends in themselves. Apparent justifications for them ('the advancement of knowledge is man's supreme activity') usually turn out to be no more than disguised assertions of their ultimate value.

If the method of polling places one's own highest and 'irreducible' goal low in the general rank order, defeat may not be easily accepted. The losers may withdraw their earlier acceptance of that method and say that there has been foul play, or they may continue covertly to do what they think is right. Or they may be openly defiant, thus rupturing the higher form of consensus—collegiality. In any case, the poll has not uncovered an underlying consensus. If they accept the goal which, according to the poll, has the highest rate of approval and the lowest rate of disapproval, then this is evidence of a consensus of a sort, but it is not about goals. The goal which they supported before the poll, and which they represented as an irreducible value, in fact has been reduced in favour of something else: prudence, fellow-feeling, the hope of getting something in return, perhaps weariness, or some combination of these and other influences. They have, in fact, allowed themselves to be persuaded and it may be that the actual function performed by counting heads, in whatever way, is not to uncover the non-existent consensus about goals and values, but only to minimize dissent and allow some dissenters to climb down without too much loss of face. The common value is not substantive but, as we noted, procedural.

CONSISTENCY AS A POLITICAL PROCESS

Evidently neither logical nor electoral manipulations are going to reveal a hidden consensus about goals, where none exist. Let us, then, shift the direction of the enquiry away from the pattern of goals towards what the people concerned think about this pattern, and towards an activity which arises out of disaccord—persuasion.

When Shils insists (1975: 5)—who is not tempted to agree?—that there is something in the centre of every society, both values and ruling authorities for whom the people have at least a 'minimum appreciation', he is not only defining what the word 'society' means, and at the same time making an assertion about the way people will in fact be found to behave, but he is also giving voice to the discomfort which human beings feel in the face of disorder—of intellectual inconsistency, directives which cancel each other out, beliefs which contradict, goals vehemently asserted which cannot even in a small degree be realized in actual life, values proclaimed irreducible which every one knows are traded off without hesitation, and in general of a failure to attach meaning to and find a pattern in what goes on around one. When we are caught out in any of these forms of inconsistency, our first effort is to find some way of denying the inconsistency, saying that it is apparent but not real. If that fails, then we are likely to find reasons why nothing can be done to remove the inconsistency, thus restoring a kind of consistency but at a higher level. (For example, in one university the faculty, in what they did, supported the education of a privileged elite but at the same time gave verbal priority to mass education through extension teaching, a clear inconsistency. They did so because the political pressures on them were equally strong from both directions, and to ignore this reality would be another form of inconsistency—aiming at goals which are unattainable.)

Why do people tolerate inconsistencies? It may be said that it is in the nature of men to do so. One is also tempted to say that people who do so do not use their intelligence or, if intelligent, are lazy. But neither of these remarks are satisfactory: they may well be true in particular cases, but they put a premature end to the enquiry. More promising explanations look at the situation of the person who holds inconsistent views. There may be psychological pressures such that if he abandons either one of the two conflicting views or goals, he is subjected to intolerable feelings of guilt. There may be political pressures, as in the forementioned example. So he suppresses whatever desire he may have for intellectual consistency and for consistency in practical reasoning, and goes on muddling through.

'Situation' may also be more widely construed. One's views on the way the institution around one is, and the way it should be, are continually tested by experience. Values, for instance, which tend to eliminate the person holding them, are self-destroying. Values given institutional support but constantly denied by experience are likely to be adjusted. This experience is not a once-for-all affair acting simultaneously all parts of the view, but consists of different events occurring at different times, each one affecting only a part of the whole picture. Discrepancies, therefore, may come about simply because one part of the map, as it is sometimes called, has not caught up with the rest. It reflects yesterday's experiences: tomorrow's experiences may set the matter straight and remove the inconsistency, providing, of course, that in the meantime there has not been another source of turbulence. According to this view, which seems sensible, one should scrutinize views on the institution not for consistency but rather examine them to understand the *process of striving for a consistency* which will probably rarely be achieved.

There is another way of approaching the problem—a way which does not negate the other answers but supplements them. One may regard each conversation as presenting not simply a view on the university and its goals, but rather as a rhetoric about that subject. These conversations—or indeed any account of a view on the institution—are not so much a tale of what a man thinks and desires but, rather, are directed at a particular audience and are intended to persuade that audience to accept a belief or adopt a goal or value. Since different audiences arrive with different presuppositions, the exposition will vary so as to accommodate these differences. Some parts of one person's stated view may be out of accord with other parts because the audiences to which they are addressed are different from one another.

This does not, of course, remove the notion that to be consistent is a good thing. In fact, if you point out that someone has said different and apparently incompatible things to different audiences, he is likely to be indignant or embarrassed and will probably go through the same arguments, as those mentioned earlier, to remove or justify the inconsistency. So at this level, too, one can look for a process of striving for consistency, or at least for its appearance and perhaps for indications of self-deception.

Just as individuals feel discomfort when they are told that their central values and beliefs are pointing in several different directions at once and they may have to choose, so also there is an institutional equivalent. This lies in the process by which the members attempt to reduce the distance between their different points of view. They do so by means of that rhetoric, which I used in the foregoing to account for apparent inconsistencies in a man's arguments.

If we put aside the idea that the various goals are best seen as part of a unified set of directives for action and policy, we can think of them instead as themes available in that culture to make assertions about what the relevant world is like and should be like. If these themes are not consistent with one another, then that is no longer a reason for surprise or embarrassment. To search behind them for some hidden consensus makes no sense, for in themselves they are just the opposite of that—occasions for conflict.

This does not eliminate the sense of discomfort which arises when it is pointed out that an institution which tries to march simultaneously in several different directions will probably get nowhere. Nor does it remove the practical necessity for making decisions or the feeling that there ought to be something at the centre. But it does suggest that whatever is at the centre, it does not have to be a set of substantive and consistent values and directives and goals which command a general consensus.

CONCLUSION

I have been in search, in this paper, of what is at the centre of universities and other institutions of higher education, of their 'dogmatic foundation'. I began with what seemed to be a plausible candidate for substantive central directing values—namely, the stated goals of the institution; and I have been looking for a consensus. This consensus was not found, and in its place I have put an arena in which conflicting goals are debated.

I have also put aside that view of values which sees them only as internalized directives for action. This rejection can very easily be misunderstood and interpreted as an easy cynicism that regard. most statements of motivation as hypocritical, the reality lying with that famous *cui bono*? ('find out who benefited'). This is not my position. Undoubtedly, people are moved by values other than material gain. But the difficulty is that it is sometimes very hard, if not impossible, to discover what actually moved someone to perform a particular action (he may not know himself), and even if one could know this, that knowledge would still leave in the dark the greater part of communication and interaction. To know what a man wants, or to be able to say what a society or an institution wants (the notion of psychological reality in the last two cases becomes far-fetched), is not the same as understanding the accepted or the effective ways in that culture of realizing one's goals.

So we put to one side the notion that in finding out what people say about the way their institution does work and should work, our first task is to find out if that is what they 'really' think. We can bracket that question away and ask instead a question which belongs to politics and communication. What image are they trying to put across? What audience do they have in mind? Why do they choose that image rather than other available ones?

We are no longer in search of a definitive order of 'values and beliefs which govern the society', but of an order of persuasive strategies by means of which various values and beliefs are advanced.

REFERENCES

BLACK, DUNCAN. 1963. *The Theory of Committees and Elections*. London: Cambridge University Press.

BLAU, PETER M. 1964. *Exchange and Power in Social Life*. New York: Wiley.

GOULDNER, ALVIN W. 1957. 'Cosmopolitans and Locals: Toward an Analysis of Latent Social Roles', Parts I–II, *Administrative Science Quarterly*, 2: 281–306, 444–80.

SHILS, EDWARD. 1975. *Center and Periphery*. Chicago: University of Chicago Press.

SINGER, MILTON (Ed.). 1973. *Entrepreneurship and Modernization of Occupational Cultures in South Asia*. Duke University: Monograph Series.

6

Administrators' Attitude to Change: An Exploratory Study

A. P. BARNABAS

INTRODUCTION

The administrative system in developing countries, by and large, has to act as an agent of social change. There have been some questions as to whether an administrative system, which is highly bureaucratic by nature, can be an effective agent of change. The elaborate hierarchy of administration leaves little scope for it not to have a rigid structure. Merton (1952) suggests that persons in a bureaucratic system tend to be less innovative and less prone to change. It has also been suggested that the Weberian model of bureaucracy is more for maintaining *status*

Acknowledgement: I wish to acknowledge the help rendered by Shanta Kohli Chandra in preparing this paper.

quo than for bringing about change. Some studies carried out at the block level in India (Mathur 1972; Roy 1975) suggest that development orientation is rather low among bureaucrats in India.

Anthony Downs, in his book *Inside Bureaucracy* (1967), suggests the following hypotheses regarding the attitudes of persons in a bureaucratic system:

1. An official is likely to become a conserver (*a*) the longer he remains in a given position, (*b*) the older he becomes, (*c*) the longer he remains within a bureau, if he is still not in the 'mainstream' of promotion to the very top, and (*d*) the more authority and responsibility he has, if he is still not in the 'mainstream' of promotion to the very top and if he has strong job security.
2. The middle level of a bureau hierarchy normally contains a higher proportion of conservers than either the lowest or the highest level.
3. The proportion of conservers among the older officials is usually higher than among the younger ones.

These hypotheses can be tested by studying either the value orientation or the specific attitudes. The basic assumption is that if the members of the bureaucracy have a positive attitude to change, they can act as effective agents of social change.

OBJECTIVES OF THE STUDY

This study has been designed, firstly, to explore the following possibilities:

1. Studying the administrators' attitudes to change;
2. replicating a study in another culture; and
3. using an indirect approach to study attitudes.

The design has been prepared with the help of some of the hypotheses suggested by Downs (1967). In particular, his hypotheses have been adapted to suit the situation in India.

The purpose of the study, more specifically, is to look at the following questions:

1. What is the general attitude of administrators to change?
2. Do age, education and number of years of service have a bearing on the attitude to change?
3. Is there a difference in the attitude to change between persons belonging to different services, like the Indian Administrative Service (IAS), Indian Police Service (IPS). Central Secretariat Service (CSS), Provincial Civil Service (PCS), and so on?

It is hypothesized that:

1. The overall attitude of the administrators to change will be negative.
2. The older a person, the less prone he will be to change.
3. The longer in service, the more likely he is to be a conserver.
4. The higher the status of a bureaucrat, the more prone to change will he be.

The two crucial concepts in the study are 'attitude' and 'change'. 'Attitude' can be defined as manner of acting representative of a feeling or opinion. It is the 'opinion' or 'feeling' that is being analyzed in the study, not the manner of acting. Attitudes can also be said to be beliefs around an object or situation predisposing one to respond in some preferential manner. Change *per se* is not being measured. The emphasis is on knowing the administrators' attitudes to change. Change can refer to innovative behaviour of the administrator himself, or the administrator's capacity to stimulate change in the behavioural patterns of citizens. However, neither of these aspects are considered in the study. The assumption is that if there is a positive attitude to change, there would be a positive inclination to bringing about change.

METHODOLOGY

The questionnaire was adopted from Neal's study (1965) of the value orientation of Catholic priests in the US. According to her, 'The results suggest that the same model can be used to examine the change process in political, behavioural, business and other social systems by simply redoing value and interest items in content relevant to the system under examination' (1965: 160–61). Some adaptations were made in the questionnaire to make it relevant to the Indian administrative system. Only those items which were likely to provide insights into the administrators' attitude to change in India were chosen.

The questionnaire consisted of 26 statements. The respondents were requested to score each item on a scale from –3 to +3. If they disagreed strongly with a statement, they had to mark –3, if they disagreed –2, and if they disagreed slightly –1. Similarly, if they agreed slightly +1, if they agreed +2, and if they agreed strongly +3. The following are some of the statements included in the study.

1. Young people sometimes get rebellious ideas, and as they grow up they ought to get over them and settle down.
2. The future is in god's hands. I will await what he sends and accept what comes as his will for me.
3. Man should try to rectify in creation everything he can rectify.
4. Any organizational structure becomes a dead weight in time and needs to be revitalized.

Each statement was graded depending on the scores given by the respondents. For example, in the first two statements, if some respondents indicated strong agreement, they were given one mark each, as both the statements are against change. On the other hand, if they disagreed strongly (that is, if they gave –3), a score of 6 was given to them. Similarly, as regards the other two statements which were pro-change, if they agreed strongly, a score of 6 was given, and if they disagreed strongly, the score was 1.

As there were 26 statements, the lowest score could have been 26 and the highest 156. However, the range was between 76 and 139. A scale was developed as follows:

Below 80	Against change
80 to 100	Neutral
Above 100	Favourable to change

THE SAMPLE

The sample for the study was drawn from among the participants in various courses conducted for administrators by the Indian Institute of Public Administration, New Delhi. Courses were chosen at random.

The questionnaire was distributed among 200 participants in various courses. Of them, 184 responded. This sample, it was assumed, would be adequate to provide sufficient data to answer the questions raised earlier.

A majority of the respondents were between 30 and 50 years of age. While 55 per cent had obtained post-graduate degrees, 43 per cent were graduates. About half (48 per cent) had more than 15 years of experience, while about 30 per cent had been in service for 11 to 15 years, and 17 per cent for 6 to 10 years. Of the respondents, 40 per cent were from the PCS, 18 per cent from the IAS, 10 per cent from the IPS, and 11 per cent from the CSS. The Railways, Post and Telegraphs, Revenue and Excise, and so on, grouped under 'Others', accounted for 21 per cent of the respondents (for details see Tables 6.1 to 6.4).

Table 6.1 ***Distribution of Respondents by Age***

Category	*Number*	*Percentage*
Below 30 years	12	7
31–40 years	76	41
41–50 years	72	39
51 + years	24	13
Total	184	100

Table 6.2 ***Distribution of Respondents by Education***

Category	*Number*	*Percentage*
Under-graduate	3	2
Graduate	80	43
Post-graduate	101	55
Total	184	100

Table 6.3 ***Distribution of Respondents by Length of Service***

Category	*Number*	*Percentage*
1–5 years	10	5.5
6–10 years	32	17.0
11–15 years	54	29.5
15 + years	88	48.0
Total	184	100.0

Table 6.4 ***Distribution of Respondents by Service Category***

Category	*Number*	*Percentage*
Indian Administrative Service	34	18
Indian Police Service	18	10
Central Secretariat Service	20	11
Provincial Civil Service	74	40
Others	38	21
Total	184	100

ANALYSIS OF DATA

The manner of scoring has already been indicated. It was found that a majority of the respondents were in favour of change. Table 6.5 shows that 71 per cent of the respondents had a positive attitude to change and only 2 per cent were against it. The remaining 27 per cent were 'neutral' (that is, according to the basis of the scoring indicated earlier, it was difficult to place them either as positive or as negative towards change). The first of the four suggested hypotheses, therefore, is not tenable.

Table 6.5 ***Distribution of Respondents by Attitude to Change***

Attitude	*Number*	*Percentage*
Against change	3	2
Neutral	50	27
For change	131	71
Total	184	100

The second hypothesis is that an official is likely to become a conserver the older he becomes. There is some indication in the data to confirm this hypothesis. Eighty-three per cent of those below 30 years and 68 per cent of those between the age of 30 and 40 were for change. In the age group 41–50, 75 per cent were for change, and 62 per cent in the age group 50+. It seems a larger percentage of respondents in the age group 41–50 are for change as compared with the percentage of respondents in the age group 31–40.

The distribution of 'neutral' respondents raises the same problem, as the percentage is higher in the age group 31–40 years than that in the age group 41–50 years. However, the percentage for those above 50 is significantly higher than those for the other three age groups (see Table 6.6).

Table 6.6 ***Distribution of Respondents by Age and Attitude to Change***

Attitude	*Age*				
	Upto 30 years	*31–40 years*	*41–50 years*	*50 + years*	*Total %*
Against change	–	3	1.5	–	2
Neutral	17	29	23.5	37.5	27
For change	83	68	75.0	62.5	71
Total	100	100	100.0	100.0	100
	(12)	(76)	(72)	(24)	(184)

Note: All figures are percentages. Figures in brackets denote *n*.

The third hypothesis is that persons who have put in longer years in service will be more conservative than those who have worked for a lesser number of years. It will be seen from Table 6.7 that this hypothesis is not confirmed by the data. Of those who had worked for less than 5 years 60 per cent were for change, while the figures for the others were 75 per cent for those who had worked for 5–10 years, 76 per cent for those who had worked for 10–15 years, and 68 per cent for those who had worked for more than 15 years.

This particular variable needs to be tested further. It is possible that those who are new to the system may be more conservative than those

Table 6.7 ***Distribution of Respondents by Attitude to Change and Years of Service***

Attitude	*Service*				
	Less than 5 Years	*5–10 Years*	*10–15 Years*	*15 Years*	*Total %*
Against change	–	–	4	1	2
Neutral	40	25	20	31	27
For change	60	75	76	68	71
Total	100	100	100	100	100
	(10)	(32)	(54)	(88)	(184)

Note: All figures are percentages. Figures in brackets denote *n*.

who have been in the system for a longer time. The latter may have a greater awareness of the problems and dysfunctionalities of the system, and hence may want to bring about change. A sense of security due to long years of service may also allow for more innovative behaviour. The data raise questions about the relationship between innovative attitudes and years of service. The assumption that the younger people with fewer years of service are inclined to change is not supported by the data.

As regards education, two variables were taken into consideration—level of education and subjects studied. A higher percentage of the post-graduates were in favour of change as compared with the graduates and under-graduates. Among the under-graduates, the percentage of positive respondents was 33, while among the graduates it was 69, and among the post-graduates 74. There is, therefore, a positive correlation between the level of education and the attitude to change.

Those who had studied science were more pro-change (81 per cent) than those who had studied Arts (66 per cent) and Commerce (73 per cent). One of the problems, however, is that there is a wide variation in the number of respondents—of the respondents 115 had studied Arts, while only 54 had studied Science and 15 Commerce (see Tables 6.8 and 6.9).

The respondents were classified according to the services to which they belonged—the IAS, IPS, CSS, PCS and 'Others' (see Table 6.10). The assumption was that those in the higher services (such as the IAS and IPS) would have a more positive attitude to change as compared with those in the CSS and PCS. The percentage of those favouring

Table 6.8 ***Distribution of Respondents by Attitude to Change and Level of Education***

Attitude	*Education*			
	Under-graduates	*Graduates*	*Post-graduates*	*Total %*
Against change	–	1	3	2
Neutral	67	30	23	27
For change	33	69	74	71
Total	100	100	100	100
	(3)	(80)	(101)	(184)

Note: All figures are percentages. Figures in brackets denote *n*.

Table 6.9 ***Distribution of Respondents by Attitude to Change and Type of Education***

Attitude	*Education*			
	Arts	*Science*	*Commerce*	*Total %*
Against change	3	–	–	2
Neutral	31	18.5	27	27
For change	66	81.5	73	71
Total	100	100	100	100
	(115)	(54)	(15)	(184)

Note: All figures are percentages. Figures in brackets denote *n*.

Table 6.10 ***Distribution of Respondents by Attitude to Change and Type of Service***

Attitude	*Service*					
	IAS	*IPS*	*CSS*	*PCS*	*Others*	*Total %*
Against change	–	–	–	1	5	2
Neutral	20	11	15	38	26	27
For change	80	89	85	61	69	71
Total	100	100	100	100	100	100
	(34)	(18)	(20)	(74)	(38)	(184)

Note: All figures are percentages. Figures in brackets denote *n*.

change was 61 among the PCS, 85 among the CSS, 80 among the IAS, 89 among the IPS and 69 among 'Others'. It is difficult to draw any specific conclusions because the number of respondents was rather small in the IPS and CSS. It would seem that, comparatively, a lower percentage of those in the PCS were for change. Overall, a much higher percentage among all groups were favourable to change, hardly any against change, and a few were neutral.

SUMMARY AND CONCLUSION

The study was exploratory in nature. The effort was to explore three areas. The first was the possibility of studying attitudes. It is often suggested that studying attitudes is not possible as the verbal expression may be different from actual behaviour. It is difficult to accept this proposition in its entirety. If such a proposition is accepted, it would then be impossible to undertake any study in the area of attitudes. This study confirms that attitudes can be studied. An indirect approach was adopted for this. Certain statements were made and the respondents were requested to indicate their agreement or disagreement. Ninety-two per cent of the sample responded to the questionnaire given to them. The respondents had given their scores for all the statements. From the responses, it was possible to draw out their attitudes. The responses indicate that this method was relevant, and evoked sufficient interest to get substantial responses.

Another area of exploration was the possibility of replicating in one culture a study conducted in another. This study indicates that it is possible to do so. However, this is not to suggest that the culture did not have an impact on the responses. It could be argued that the responses in regard to the statements had a cultural bias. One of the statements was: 'The future is in God's hands. I will await what He sends and accept what comes as His will for me.' A high percentage of the respondents agreed with the statement. It is possible that, in most traditional societies, the responses to such a statement would be strongly favourable.

While a few statements refer to religion, the majority of the statements did not have any such discernible bias. For example, one of the statements was: 'My first reaction when I think of the future is to be aware of its danger.' It did not have the same type of value bias as the previous one did. The attitudes of people are based partly on internalized values. These values manifest themselves in people's reaction to the statements. For example, an agreement with the statement regarding the future being in god's hands indicated a negative attitude to change. Of the 26 statements given, 12 were positive to change and 14 negative. Informal discussions with the respondents indicated that they were not aware of the implications of their agreeing or disagreeing with a given statement. Such an awareness could have biased the responses.

It would thus follow that while studies in one culture can be replicated in another, cultural factors have to be given due consideration in the analysis of data.

An indirect approach seems to be appropriate and relevant in studying attitudes. Possibly, the methods of scoring can be refined. However, this does not deny the validity of the method in studying attitudes.

This exploratory study indicates that (*a*) the majority of the officials were in favour of change; (*b*) there is high correlation between the level of education and attitude to change; (*c*) there is no clear correlation between the attitude to change and the years of service, though officials with fewer years of service seemed to be more conservative; and (*d*) the data are not adequate to arrive at any conclusion regarding the type of service and the attitude to change.

This study has shown that the particular methodology it uses does enable one to study the attitudes of administrators. In future studies, greater attention needs to be paid to the sample so that the various hypotheses can be tested more rigorously. It may be more appropriate and functional to study officials from the ministries actively engaged in bringing about change (for instance, Agriculture, Rural Development, Family Planning, and Social Welfare). The importance of knowing the administrators' attitudes to change in a developing society need hardly be emphasized because the administrative system, by and large, in most of the developing societies is an agent of change.

REFERENCES

DOWNS, ANTHONY. 1967. *Inside Bureaucracy*. Boston: Little Brown.

HIGINBOTHAM, STANLEY J. 1975. *Cultures in Conflict: The Four Faces of Indian Bureaucracy*. New York: Columbia University Press.

MATHUR, KULDEEP. 1972. *Bureaucratic Response to Development: A Study of Block Development Officers in Rajasthan and U.P.* Delhi: National Publishing House.

MERTON, ROBERT K. (Ed.). 1952. *Reader in Bureaucracy*. Glencoe: Free Press.

NEAL, MARIE AUGUSTA. 1965. *Values and Interests in Social Change*. New York: Prentice Hall.

ROY, RAMASHRAY. 1975. *Bureaucracy and Development: The Case of Indian Agriculture*. Delhi: Manas.

7

Social Space and Social Interaction in Pune City

Y. B. DAMLE

SOCIAL SPACE AND SOCIAL INTERACTION: SYMBOLIC AND SUBSTANTIVE

Ecologists have rightly stressed the relevance of physical space for social relations. Social space, however, is of direct importance for the study of social structure as well as change. By social space is meant not only the living together of people sharing common circumstances but also their position in a scheme of societal placement, hierarchy, gradation, and so on. In terms of occupying a particular physical space, the sharing of various facilities and amenities is also implied, but there are also constraints arising out of such sharing. Since people are located in social spaces and identified with particular social spaces, it becomes possible to distinguish between groups of people occupying different social spaces. While physical space primarily accounts for social space, there need not be complete correspondence between physical and social space. For example, a fashionable area in a city may be encumbered with a physical and social space incongruent with it (like a slum located next to it). All the same, when any town or city grows, efforts are made to establish near correspondence between physical and social space, in the sense that those who occupy the same physical space also tend to occupy more or less the same social space.

A given social space almost simulates a primary group situation, because of the shared facilities, ethos and values of those occupying it. Social interaction flowing out of such sharing is characterized by intimacy. It is also both symbolic and substantive. In its symbolic aspect the entire gamut of culture and values is reflected, while in its substantive aspect the utilitarian dimension is taken care of. Of course, such a distinction between the symbolic and substantive is made for analytical purposes. All kinds of relationships ranging from the most intimate to those of a relatively secondary type are subsumed under social interaction. While family, kinship and caste play a significant role in social interaction, a sense of relatedness to others, both positive and negative, is also reflected in it. In short, the beginning of the feeling of 'we' and 'they' can be ascribed largely to the relevant social space and interaction.

This paper deals with social space and social interaction primarily with reference to Pune city and its expansion across the Mutha river in the west.[1] The sociography of a city provides insights into the understanding of its organic development, in the sense that the characteristics of historical, physical and social space undergo a change in response to (*a*) the needs of the situation, and (*b*) the changing system of ideas, ideologies and values. What is handed down from the past is changed, not completely discarded. Organic development, rather than development from without, needs to be properly analyzed, understood and appreciated instead of getting caught in the polarity of tradition and modernity, since the latter does not adequately represent the complex reality.

In the case of Pune city and its expansion across the river, the acquisition of modern education, entrance into the secular occupa-

[1] The material used in this paper is drawn from Jayashri Gajanan Joshi's, 'Different Residential Patterns and Social Interaction with Reference to Pune City', unpublished Doctoral dissertation submitted to the SNDT Women's University, Bombay, 1985. The dissertation was prepared under my supervision and originally written in Marathi.

tional structure, and exposure to new ideas, ideologies and values did not signify a mechanical, not to say senseless, rejection of what existed, but selective acceptance. This was facilitated by the nationalist and social reform movements—pertaining to the upper caste and class as well as the deprived and unprivileged, and calling for a radical transformation of the social structure. The point to be emphasized is that the inextricable intertwining of organic needs and development is much more important than purely mechanical imitation in behaviour and interaction. If the latter is emphasized, one would end up with superficial understanding. As discussed later, the emergence of bungalows, their ownership by the middle class, and the refusal by this class to imitate the earlier pro-Western and pro-British style of living provide a case of organic development—change without burning the boats, as they say.

THE VARIABLE OF TIME

Any city which boasts of a glorious historical tradition has to be studied in the context of such a tradition. People continuously refer to it because it is rooted in their consciousness. The physical as well as social spaces have been, in no small way, determined by the forces of history. The various structures, both physical and social, can be understood better with appropriate knowledge of historical tradition. In the case of Pune city, Shivaji's name is inextricably linked with its development. Maratha rule, as administered mainly by the Peshwas located in Pune, has added to the continuity of its historical tradition.

The centre of power in the Maratha kingdom shifted to Pune since the Peshwas chose to stay there and develop the city according to their needs, aspirations and ideas. The construction of *wadas* (large residential complexes), each composed of many rooms and other units to cater to various needs, both formal and informal, private and public, typified such a development. The size of a *wada*, a huge complex consisting of squares surrounded by rooms, drawing halls and so on, characterized the prosperity of a *wada*. Such *wadas* were owned by powerful *sardars* (officers in the Peshwa durbar). *Wadas* were self-sufficient, analogous to the British castles, since they were well protected and provided with all amenities and facilities for more or less self-sufficient living. Moreover, each *wada* was for the exclusive use of a family, generally spanning three to four generations and, in the course of time, there was a proliferation of kin in a *wada*. Among the *sardars* there were Brahmins as well as non-Brahmins, and each *wada* was constructed according to the status of the *sardar* possessing it. In fact, the status of a *sardar* could be gauged by the kind of *wada* he had constructed. Similarly, the prosperity of the ruler was reflected in the construction and design of these *wadas*. They served as residences for mini-rulers, with all their pomp, grandeur, power and affluence.

While many of the *wadas* continued to exist during British rule, their grandeur, prosperity and power diminished considerably, if not vanished altogether, over time. The transition from *wadas* for self-use to *wadas* for renting reflected their and their owners' deteriorating condition. Moreover, in the wake of mounting poverty and unemployment, structures called *chawls* (tenements) emerged to cater to the needs of the poorer sections of the population. While the owners and occupants of *wadas* symbolized higher income and status, the occupants of *chawls* symbolized lower income and status. Even the owners of *chawls* were not on the same level as those of *wadas*.

Due to British influence, the educated and modernized sections of the population, such as government officers and professionals, were attracted by the bungalow-type construction of the British—a case of reference group behaviour. This section of the population felt that in order to enjoy status and prestige under the new dispensation, that is, British rule, they could construct and live in bungalows which had become the hallmark of modern living. Modern education, employment in the higher government echelons, and new laws (such as the Gains of Learning Act) went hand-in-hand with living in bungalows. The highly educated persons were both conscious and watchful of their independence and authority. In *wadas*, on the other hand, the supreme authority of the eldest male head of the family continued undisturbed. This change in the pattern of authority, decision-making, status and prestige in favour of those who were highly educated and were either employed in prestigeous positions in the government or worked as professionals, affected social interaction both within and outside the *wada*.

The owners of bungalows were necessarily few in number, but their style of living and ideas about individual freedom affected those who were educated but not so well placed. The latter were led to construct 'apartments'. The occupants of apartments were away from *wadas* and bungalows in terms of income, status, prestige, authority and power. Living in apartments symbolized their quest for freedom from traditional constraints, such as those of kinship and caste, but without the financial ability to completely steer clear of these constraints. Even then, considerable modification appeared in their life-style, relationship between the two sexes, education of women, and employment.

The First World War provided an impetus to the construction of *chawls* and the utilization of *wadas* for rental. Between the First and the Second World War, the condition of *wadas* deteriorated further due to changes in property laws, emphasis on partitioning of *wada* property by coparceners, and consequent changes in the relationship between owners and tenants. *Pari pasu*, with this development, the proliferation of *chawls* as well as apartments took place, which further affected the position of the owners as well as the occupants of *wadas*. Increasing economic pressure and the enhancement of higher education also meant different perceptions of life, particularly of its future prospects, by different generations in the same family. This had important implications for splitting up the joint family and women's status.

Wadas provided sufficient privacy for holding meetings to discuss plans and programmes connected with the independence movement. Both Brahmin and non-Brahmin *wadas* were intimately connected with the movement. For example, Tilak and Jedhe, two prominent nationalist leaders, owned *wadas*, which became active centres of the movement. The British government conducted raids on them to search for incriminating papers. There is sufficient evidence of how neighbouring *wadas* helped to transfer the papers to them in good time. This was possible because *wadas*, though large and complex, were not isolated from each other. During the atrocities committed by the British and their relentless search for persons connected with political action, particularly the terrorists, the latter enjoyed the support of the *wada* occupants. The family members of those who were sent to jail were also looked after by the *wada* occupants. *Wadas* thus played a political role as well.

Chawls and their residents played a much smaller role in politics because *chawls* offered hardly any privacy to their occupants, and many of the residents were petty government servants and therefore could not afford to associate themselves with political activity. As one can easily imagine, the bungalows and their residents played hardly any role in the independence movement, because the heads of families residing in them were high government officers or professionals. On the other hand, those staying in apartments could provide the required privacy for political discussions and active involvement.

The various residential structures mentioned in the foregoing and their occupants were influenced by social reforms initiated by Phule, Agarkar, Karve, Shinde and others. The impact of these social reformers and their radical thinking, however, was different in the four different structures (namely, *wadas*, *chawls*, bungalows and apartments). For example, the atmosphere in *wadas* was not at all conducive for the ready acceptance of social reforms pertaining to women's status, education and employment, a radical stance against the caste system in its entirety, and rationality and secularism. Nevertheless, the efforts of social reformers affected the inhabitants of all the structures in the course of time. This is reflected in the change in social interaction, which will be discussed later.

Due to this change, the concern for women's education, employment and status took firm root in Pune, as exemplified by Phule's and later Karve's efforts for women's education. There was, however, criticism that social reforms pertaining to women's education, employment and status were confined mainly to the educated middle class. Women's education and status were of great interest to the inhabitants of bungalows also, which were more or less emancipated from the hold of the authority of the joint family. The ideas of nationalism, secularism, freedom, equality, social justice and human dignity also made inroads into the various structures and their inhabitants, although there tended to be a selective acceptance of these ideas. For example, the non-Brahmin movement derived its support from non-Brahmin *wadas* and *chawls*.

IDEAL TYPE CONSTRUCTION

To appreciate properly the distinctive features of the spatial structures, and to be able to understand the sense, spirit, ethos and pattern of interaction in them, I shall use the method of ideal types. The construction of an ideal type is sometimes conceived of in static terms. However, because of the historicity of tradition, in the present case, the ideal types render the possible depiction of organic unity over a period of time. Both symbolic and substantive interaction are taken care of by the emphasis on spirit and ethos as well as on the structure of relationships and interaction. The people residing in these different kinds of spatial structures necessitated as well as afforded a certain pattern of interaction among them, both within and outside the structure. The physical location of various structures, their social spaces and the dimension of time can be properly analyzed with the help of ideal types. A description of these types follows.

WADAS

As mentioned in the foregoing, *wadas* were rooted in historical tradition and reflected a particular structure of interaction. Their physical construction, including the huge complex of rooms and other units, and their self-sufficiency have been conducive to maintaining their pristine purity, with the least interference from outside. Their various units and sub-units bear testimony to this. There were units for the exclusive use of women, men, formal occasions, entertainment, religious rituals and ceremonies, and so on. *Wadas* provided ample space for children to play. The unity of generations in the family meant that children did not have to seek playmates from outside the *wada*, since there were numerous brothers, sisters, cousins, and so on, within the *wada*.

Wadas were governed by the caste system. It has already been mentioned how there were Brahmin and non-Brahmin *wadas*. There was a hierarchical relationship among them on the basis of caste, economic affluence, education, and so on.

By and large, women were condemned to an inferior position and had to submit to absolute domination as soon as they entered the conjugal household after marriage. The age at marriage ranged from 3 to 12 years, although this changed in the early twentieth century. The education of women had not advanced sufficiently to improve their status. The system of marriage, choice of spouse, and marriage rituals and ceremonies reflected the inferior position accorded to women. The relationship between the husband's and wife's relatives was hypergamous in nature. Utmost importance was attached to the male child. A woman's failure to give birth to a male child could result in her husband bringing in a co-wife. The position of the widow and the woman discarded by her husband was particularly pitiable.

Wadas were suitably guarded. Their construction facilitated the protection of their inhabitants, particularly of women and children. Life in the *wadas* centred around rituals, ceremonies, fasting, scrupulous insistence on rules of diet and dress, and conjugal relations. A young husband and wife were not allowed to look into each other's eyes in the presence of elders, leave aside talking to each other and expressing their intimacy directly or indirectly. Scrupulous care was taken to maintain detailed accounts and control over monetary resources, which was reflected in the budgeting pattern for various items. Men had access to education, while women were not allowed to be educated. Children were groomed for traditional religious education, which ensured their socialization into the family.

It has already been mentioned how *wadas* provided protection to political discussions and political activity. This holds good primarily for those *wadas* which were essentially for the self-use of owners. Even in *wadas* where parts were rented, the forces of tradition, such as caste (and sub-caste), prevailed and the supreme authority of the *wada* owner continued to dominate. For all practical purposes, the owner both behaved and was accepted as the head of the joint family. As in the case of the *wadas* for self-use, all kinds of restrictions regarding diet, dress, personal interaction within the family, relationship between the sexes and social intercourse were imposed on the tenants.

I remember how, even in the late thirties, life in a *wada* more or less conformed to the pattern mentioned in the foregoing. I used to visit some of my relatives residing in *wadas* in Pune occasionally, and personally experienced this atmosphere and the attitude of deference towards the *wada* owner. The most redeeming feature of this kind of relationship was the intense cooperation both among the tenants and between them and the landlord in respect of mutual help of every kind.

CHAWLS

All the tenements in a *chawl* were located in a row, with a common verandah and common toilet facilities at a distance. In the adjoining space, there was a water tap and facilities for washing utensils and clothes. The occupants of tenements could not enjoy the privacy enjoyed by the occupants of *wadas*, since the former belonged to different families and had to share the public facilities, the common entrance and verandah, and the utilities.

The occupancy of *chawls* conformed with the owner's caste and even sub-caste. There were *chawls* owned by Brahmins as well as non-Brahmins. The occupants belonged to the lower economic strata and were generally employed in relatively petty jobs, which added to their sense of inferiority vis-à-vis the owner. The owner lived in a different and much better residence, and therefore had no day-to-day contact with the occupants. It was only at the end of the month that he would go there to collect the rent, but preferred to send his agent to do the job.

In keeping with the layout of the *chawl*, the lack of privacy characterized the personal relationships in it. This was, however, compensated by intense inter-tenant cooperation, mutual understanding and goodwill, and sharing in fortunes as well as misfortunes. The occupants of a tenement identified with the *chawl* and each other, and as such provided necessary protection and support to each other. There was a great deal of camaraderie among them. Because of the common denominator of caste, there was a cultural homogeneity in respect of diet, dress and observance of rituals and ceremonies.

The nature of the family varied with the space available. Even if the occupant's parents were also living with him, the authority rested more or less with the earning member. This is not to suggest that the parents were less respected. The rigid regimen operating in the *wadas* could not function in the *chawls*, partly due to the lack of privacy. In the *chawls* neighbours could always run to the rescue of an oppressed daughter-in-law, unlike in *wadas* where she was condemned to a life of inferiority and suppression. In *chawls*, after the husbands left for work, women could get together and indulge in small talk and cooperative activities, such as preparing pickles and *papads*. Due to the scarcity of physical space in *chawls*, a great deal of premium was put on inter-tenant cooperation for the successful carrying out of various ceremonies, like naming, initiation, marriage and death. Here the children had plenty of playmates who were not related unlike *wadas*. As mentioned earlier, the identity of a chawl was reflected in its children's play groups.

In keeping with the socio-economic status of the tenants, the level of education was fairly low. Mainly the head of the family acquired education so as to qualify for a job, while the wives did not possess that level of education. Daughters of the family were, however, allowed to go to school—primary as well as secondary. The age at marriage being low, though higher than that in the *wadas*, there were limitations on the level of daughters' education. Sons, on the other hand, had to be educated so as to qualify for a job, preferably in a government office or as a teacher in a primary or secondary school.

Generally, sons would follow the father's occupation, and therefore there was hardly any difference in their style or perception of life or in their aspirations as compared to their fathers. In the selection of a spouse, similar socio-economic and cultural background was stressed, so that the pattern of the family would not be disrupted. As for religious practices, tradition reigned supreme, but rituals could not be performed with the same degree of pomp due to the lack of facilities, as in *wadas*. Some religious festivals, like Ganesh Puja, had become a common activity for the entire *chawl*.

On account of the nature of their employment, the occupants of *chawls* could not participate directly in political activities, although discussions about political events (like the nationalist movement) took place more or less freely after the men returned home from work. In the non-Brahmin *chawls* in particular, heated discussions would take place about the non-Brahmin movement. The level of education of the head of the family in the non-Brahmin *chawls* differed from that in the Brahmin *chawls* because, in the case of the former, the head of the family was employed in some trade or craft rather than in a white-collar job. Inevitably, this was reflected in the education of wives and daughters.

BUNGALOWS

Since bungalows were constructed to imitate the British rulers, they meant a demarcation between their occupants and the rest of the people. The owner of a bungalow, unlike that of a *wada*, was not only educated but also occupied a prestigious position in the occupational hierarchy, primarily due to his being placed fairly high in government service. The bungalow was not only spacious but also had a separate compound. In keeping with the desire to imitate the British rulers, emphasis was placed on sanitation, hygiene and beautification. The bungalow provided adequate space to every member of the family, which was required by the higher education the member was receiving. Social space relating to bungalows was quite distinctive and higher than that of the rest of the people in the city. This was partly responsible for the relative alienation between the occupants of bungalows and the rest of the people. The independence from public scrutiny implied a relative departure from the traditional rigorous restrictions on social intercourse. Persons of lower caste were admitted into bungalows, which was unthinkable in *wadas*. There were no restrictions on the caste or religion of the visitors, particularly for formal purposes, which existed in *wadas*, and even in *chawls* by default.

Life in bungalows was characterized by relative independence from the requirements of the traditional social structure. This was clearly reflected in the position of the women of the house. Since the bungalow was constructed mainly out of the earnings of a highly educated and employed person, the balance of authority swung in his favour, which provided higher status to his wife, unlike in the case of the *wada*. Even when parents were living with the son, the authority of decision-making rested with the son and his wife rather than with the old parents. Many instances have been cited where the old parents were reluctant to shift to the bungalow from the *wada* mainly for the reasons mentioned in the foregoing.

Because of the relative freedom enjoyed by the lady of the house, she spent greater time in accomplishments and embellishments in the fields of art, craft and entertainment. She was increasingly accepted as a partner rather than as an inferior entity. The age at marriage had increased both for boys and girls, which had important implications for their future prospects, perception of life, aspirations, and so on. Decisions about marriage depended more on social, economic, educational, cultural and occupational equivalence than on anything else, although caste restrictions were observed.

Life in the bungalow was dominated by the nuclear family. Even if there were other relatives, they were accepted out of obligation, not out of right. Religious life was much less rigorous than in the *wada*. Utmost importance was attached to higher education and occupational placement, which only guaranteed the continuation of higher status, if not its improvement. Inevitably, this meant a clear-cut distinction between the relatives who were highly educated and employed and the others who were not. The concept of 'poor cousins' could be applied to the latter.

One could easily imagine a lack of involvement in political affairs among the bungalow dwellers, particularly to safeguard their official position and prospects. There was ready acceptance of reforms in respect of women's education, the relationship between husband and wife, importance of education for daughters, reducing the rigours of caste restrictions, if not altogether doing away with them, and the development of secondary groups as a result of clear-cut distinctions between formal and informal personal relationships. There was evolution in favour of achievement over ascription.

APARTMENTS

As bungalows were intended to imitate the foreign rulers in style of life, aloofness, independence, and so on, apartments signified a desire to express identity on the part of the educated middle class, who neither felt comfortable in *wadas* nor could afford to have bungalows. Apartments are proverbially a middle class necessity and typify the middle class phenomenon. Compared with the inhabitants of *wadas* and *chawls*, the occupants of apartments were much more educated and involved in various kinds of occupations—mainly the services and professions, such as law, medicine and teaching. Even if the apartments were rented out, the owners belonged to the middle class. There was not much difference in the status, income, education or occupation between the owners and the tenants. The domination of the tenant by the owner which existed in *wadas* did not exist in apartments.

As regards physical structure, each apartment was self-contained and therefore independent, since all the facilities and utilities were provided within it. Therefore, there was no need to share any such facility or utility with anyone else. The size of the apartment was reckoned in terms of the number of bedrooms. It varied from one bedroom to two bedrooms or more. In addition, there would be a kitchen, a dining room and a drawing room.

The apartments were usually located in newly developed areas away from the main city, which facilitated the emergence of a new culture, ethos, life-style and aspirations. While bungalows were separated from one another, apartments were found in clusters or complexes and therefore did not have to face the problem of security.

The level of education of women in apartments, as distinguished from those in *wadas* was considerably higher than, and in some cases almost equal to, that of men. This had significant implications for the relationship between husband and wife and for women's employment and participation in decision-making. The nuclear family pattern prevailed in apartments like in bungalows. The liberation of women from traditional constraints meant a new pattern of family interaction. The occupants of apartments, being both educated and independent, were free to pursue their interests, whether aesthetic, artistic or recreational.

Among the dwellers of apartments, the adherence to traditional religious rituals and ceremonies had considerably receded, although women, however educated, did not eschew religion and the attendant worship or observance of certain regulations. The dietary restrictions of caste tended to evaporate among them, while the entertainment of friends and guests from different castes but belonging to the same educational, economic, occupational and cultural background increased. In keeping with the physical layout of apartments, the traditional taboo on inter-caste residential proximity was given the go-bye.

Since among the occupants of apartments both husband and wife were themselves educated, supreme importance was attached to the education of children. There was therefore status maintenance, if not status improvement, of these children. More often than not, they—both sons and daughters—tended to go in for educational careers which were both demanding and prestigious from the point of view of absorption in the occupational structure. A spirit of individualism, independence and autonomy was nurtured in the children, which resulted in the setting up of independent families after marriage.

The occupants of apartments were greatly exposed to new ideas and ideologies and not constrained by traditional beliefs and orthodoxies. They, particularly their sons and daughters, also enjoyed considerable freedom with regard to political views and ideologies. This promoted their political participation.

On the whole, apartments were somewhere between bungalows and *wadas*, trying to integrate the best of both these structures. They were considerably different from the occupants of *chawls* in education, occupation, income and life-style.

DEPARTURES FROM IDEAL TYPES AND THE NATURE OF SOCIAL CHANGE

The construction of ideal types of the four physical and social structures facilitates a deeper understanding of the spirit and substance of these structures and the life of their occupants in respect of many dimensions, such as family, marriage, religion, tradition, education, occupation, exposure to a different set of values, and so on. It may be argued that the present construction of ideal types provides a rather exaggerated picture. Indeed, there is no denying the fact that some of these structures and their occupants did exemplify the operation of the important constituents of the ideal type in its purity, and that any departure from the ideal type was looked upon as a lacuna or deviance. However, to plot social change it is necessary to resort to the construction of ideal types. It provides a meaningful base-line for analyzing both the quantum and quality of change.

In the case of the development of different physical structures and the ethos, values and patterns of interaction of their occupants across the river in Pune city, the construction of ideal types provides a significant point of reference against which quantitative and qualitative change can be recorded and understood. The four different structures and the ensuing patterns of interaction represent four different models of living, and also provide an insight into the significant role played by education, income, age at marriage, education of women, family interaction, interaction among neighbours, and the impact of new ideas and ideologies like westernization and radical political ideology.

The forementioned ideal types do not imply closed watertight systems because there have been several instances of transition from one structure to another, exemplified particularly by the transition from *wadas* and *chawls* to bungalows and apartments. Reference group behaviour, the quest for mobility, and emancipation from traditional constraints are reflected in such transition. The four models of living also take care of the diachronic dimension of social reality, since the dimension of time is organically introduced and taken care of.

In spite of the construction of ideal types, certain foci of potential and latent change have already been indicated, like the renting out of *wadas* to others, the increasing importance of higher education and government jobs (with the attendant transfers and emergence of separate households), various social reform movements, the impact of the two World Wars on the economy, society and polity, and the emergence of new ideologies. These ideologies concern not only politics but also interpersonal and inter-group relations—particularly the ideologies opposing the basis of such relationships (namely, the hierarchy founded on age, sex and caste, the patriarchal authority, and the joint family system). Similarly, the spread of education among men and women in different castes, the opportunities for employment, and the aspirations for achieving status in an otherwise caste-ridden society have substantially influenced the patterns of interaction in the physical and social structures mentioned in the foregoing. Each structure of habitation was exposed to various forces, both from within and outside. Without such a double contingency, the pattern of interaction would not have undergone the degree of change that it has in the various structures.

WADAS

Every *wada* emerged essentially for the self-use of a joint family, which enjoyed the unity of three or four generations and was more or less self-sufficient. It epitomized the traditional pattern of Hindu patriarchal authority in its pristine purity, with the scrupulous observance of authority related to age and sex. The authority of the head of the family was supreme and unquestioned. However, the introduction of modern education and secular jobs created a ferment against it. The movement for women's education also gave rise to a concern on the part of educated husbands to impart at least literacy, if not education, to women, although rarely did such husbands also want their wives to be exposed to what was happening in society by attending lectures, discussions, and so on. Thus, murmurs started against the authoritarian nature of the family and the accompanying pattern of interaction both within and outside it.

As a result of modern education, contact with persons belonging to different castes, sects and religions was initiated, which also meant the questioning of the traditional pattern of interaction imposed by and operating in the social space of the *wada*. When some parts of a *wada* were rented out, it introduced a diversity, however small, due to the heterogeneous composition of the tenants. Compared with the owners, the tenants were well-placed in the occupational hierarchy, and influenced by social and political reforms, including political movements. Inevitably, the unquestioning authority of the *wada* owners was not accepted by the tenants. This provided a leeway for social change. This process of change was heightened when the *wadas* were divided among coparceners in the course of time, so that the structure of the *wadas* and the pattern of interaction in them underwent change, leading to a change in the traditional pattern of cooperation and domination. In the *wadas* themselves, relationships tended to be much more restricted and, therefore, much more amenable to forces from outside.

One important consequence was the spectacular rise in the number of nuclear families after the division and reconstruction of *wadas*. This meant the questioning of the pattern of familial relationships based on men's supreme authority. The education of daughters, if not of wives, enhanced this change, and the education of sons for secular jobs carried it further. The size of the family also became smaller, consistent with the resources and responsibilities of nuclear families. During the last fifteen years or so, *wadas* have been reconstructed so as to transfer their ownership to the long established tenants. This has further affected social relationships because, among the new owners, a relaxation of caste restrictions has come about. Similarly, there has been a great deal of change in the number of earning women in *wadas*, particularly in Brahmin *wadas* where earlier it was unthinkable for the women of the house to work outside. In the non-Brahmin *wadas*, the women of the tenant households had already been working outside the home as domestic servants to earn some extra money. This has affected the degree of cooperation and mutual help in *wadas*.

The observance of religious rituals and ceremonies has changed with the transformation mentioned in the foregoing, because the time and resources at one's disposal have receded considerably. As mentioned earlier, the daughters of the family are also educated now, and every effort is made to provide even higher education to them. There is hardly any change in the observance of caste endogamy, with very few exceptions in Brahmin *wadas* and only one in non-Brahmin *wadas*. Marriage ceremonies have been pruned due to limitations of money, time, space and new ideas. The age at marriage of daughters has been enhanced. Strangely, with the employment of men in secular jobs, there has been greater consciousness among them of their worth, which is reflected in their demands from the wife's side at the time of marriage. As a result of the increasing education of sons, the younger generation considers life in the *wadas* inconvenient and distasteful, and therefore they aspire to go in for better accommodation as soon as they can afford it. Similarly, the influence of consumerism has given rise to a hiatus between the older and the younger generation. The enhancement of education, even undergraduate and technical education, has added to this hiatus. The significant rise in the number of institutions of higher learning in Pune in the period 1960 to 1980 testifies to the importance of higher education. This education has also meant a change in the attitude towards different castes. As already mentioned, the occupants of *wadas* have been influenced by political movements and agitations, which is reflected in their consciousness at the time of elections. In short, there have been departures in several ways from the pattern of interaction as described in the ideal type.

CHAWLS

With the changing economic circumstances there have been changes in the ownership of *chawls*. This has affected, though to a small extent, the social composition of the tenants. Pro-tenant laws have affected the relationship between owners and tenants. Pecuniary circumstances have soured the relationship between them.

The enhancement of women's education has transformed the family by encouraging educated women to go in for employment. This has helped them enhance their standard of living, reflected in the purchase of items like radios, televisions, refrigerators, and so on, although such purchasers may be very few. As a result of women working outside, the process of cooperation among the tenants is reinforced because those who stay at home are expected to perform a greater protective role in regard to the children of the women who work outside.

On the other hand, if sons and daughters acquire higher positions by way of higher education than those of their parents, their rising expectations create a distaste in them for life in the *chawls*. They search for better accommodation, which inevitably leads to the setting up of new households. Thus, families in *chawls*, which were already mainly nuclear, become even more so in the course of time.

The traditional pattern of cooperative activities within the *chawl* has undergone change. There is now a greater awareness of the importance of privacy. Even in the past, respect was accorded to higher education and employment, but it did not affect the process of integration within the *chawl*, as it has today. The occupants of *chawls* have been exposed to the emerging political ideologies, such as Socialism and Marxism, which is reflected in their voting behaviour. They also express their political opinions and views, disregarding differences of generation.

BUNGALOWS

As depicted in the ideal type, formerly bungalows were confined to certain localities nearer the city. Today, however, bungalows are constructed at a considerable distance from the city, which has in itself affected the pattern of interaction among their residents. As mentioned in the ideal type, bungalows typify more or less isolated and independent existence and minimum contact with neighbours physically removed by distance. Today, the bungalows constructed by new aspirants staying away from *wadas* and apartments belong to the middle class. They have limited resources for purchasing a larger space for constructing bungalows. This limitation is reflected in the size of the plot, the type and quality of construction, and so on. Unlike the earlier owners of bungalows, the present ones had to invest practically all their lifelong savings in constructing them, mainly due to the rising land prices and construction costs. Compared with the owners of the earlier bungalows, the owners of the present bungalows are not so affluent. The earlier owners always insisted on their independence, asserted their superiority, and kept a distance between themselves and the others.

As mentioned earlier, in the past the owners of bungalows could not completely shake off ideas about interaction between different castes. Today they are able to do so because social interaction is now guided mainly by the higher levels of education, occupation, income, and life-style. Thus, achieved rather than ascribed status has acquired greater importance. Education, ability, personality development and the dynamics of growth are greatly emphasized. There is therefore a diminution of religious faith, particularly among the youth.

Unlike women in bungalows depicted in the ideal type, women in new bungalows expect a much higher share in employment. As of today, the problem of security has increased greatly, which has affected the pattern of interaction between the occupants of bungalows and their neighbours. The pattern of marriage has undergone change due to near equality between the higher education of males and females, the participation of women in higher employment, the perception of life, and aspirations. A much higher premium is placed on professional and technical education, which is essentially competitive. There is a difference between the intra-familial interaction between the earlier and the new bungalows—the occupants of the latter are much less westernized. The organic development of bungalows, their ethos and the attendant pattern of social interaction show the artificiality of the polarity of tradition and modernity. There are also differences in maintaining relationships with other relatives, though families residing in new bungalows are also nuclear. As mentioned earlier, caste restrictions are considerably reduced. Traditional taboos in match-making are ignored by sons and daughters if there is a parity in education, employment, status and life-style. Limiting the family size is stressed.

Due to the importance attached to secondary groups, social interaction in bungalows is more by choice than obligation. The atmosphere of informality is rendered possible due to higher education, employment and life-style. Seeking entertainment among equals facilitates the process of informal and convivial living. The owner of a bungalow is asserting his independence, individuality and difference of outlook and taste in respect of his associates and friends. In keeping with aloofness from political agitations and movements in the pre-independence period, the occupants of the old as well as the new bungalows even today show political indifference, except in voting behaviour. This can be ascribed to the rising level of aspirations and competitiveness and to the lure of going abroad and seeking prestigious executive jobs and professional careers.

APARTMENTS

Unlike apartments available mainly on rent in the past, 'ownership apartments' have emerged during the past twenty-five years or so. The owners of these apartments have nothing to do with the owners of *wadas*, *chawls* and rented apartments. Ownership apartments are essentially a middle class phenomenon, where the owner has made the grade by investing his savings and getting loans from banks and cooperative societies in order to possess an apartment of his own.

As distinguished from *wadas*, there is heterogeneity among the occupants of apartments because one owns an apartment due to one's capacity to own it rather than due to caste or sub-caste considerations. The caste background of these apartment owners therefore varies from one area in the city to another. Some of the areas are regarded as prestigious because the apartments there are owned by the upper middle class, while apartments situated far away from the city are cheaper and therefore owned by the lower middle class.

In these apartments, a common denominator is provided by education, occupation, income and life-style. This has made it possible for people from different castes to stay together, though there is hardly any example of people belonging to different religions staying together. Limitations of time and physical facilities have affected the performance of religious rituals and ceremonies. Similarly, the owners of these apartments have limited resources, which limits their desire to imitate affluent society. Quite a few women are employed. The relationships within the family have undergone change with improvement in the status of women. Living in apartments has facilitated the required degree of security. In keeping with the changes, the performance of marriage ceremonies has undergone a change.

Even though there is an awareness of social injustice, the concern for one's own prosperity does not usually allow the owners of such apartments to actively participate in any political action. However, resentment against the present system is expressed by refusing to vote in the elections. Thus, frustration with the state of public affairs, particularly with corruption and political chicanery, is expressed by depoliticization.

CONCLUSION

The description of social space and interaction, as exemplified by the different residential patterns in Pune city and across the river, has brought out the importance of factors like education, occupation, age at marriage, choice of partner, ideas about living, ideologies, values, and so on. The dimension of time is reflected in the organic changes that have taken place during the process of evolution of new physical structures, the physical arrangements having consequences for social interaction. Social interaction, resulting from the various additions and modifications, suggests an organic nature of change and therefore establishes the capacity of the structure to internalize and digest changes without disrupting it. Such sociographic studies, particularly if the dimension of time is properly taken note of, would be of both theoretical and practical importance.

8

Commercialization of the Economy, and Change and Continuity in Social Structure in a Small Town in Punjab

VICTOR S. D'SOUZA
RAJESH GILL

Social structure, for this paper, is broadly defined as the interrelationships among roles and positions as well as the principles which institutionalize such relationships. The social structure of traditional Indian society is characterized by the caste system. In this system, while the members of a given caste group derive their membership by birth, they have similar roles and positions corresponding to the position and role of their caste. On the other hand, members of different castes have different roles and positions. The various castes are integrated in a hierarchy of roles and positions.

Insofar as social stratification is based on the division of labour in society, the roles and positions of individuals and groups in the caste system are, in the final analysis, derived from their occupations. Thus the caste system is intimately linked with the economic organization of the society. The element which links the economy with the social structure (that is, the caste system) is occupational prestige. The main prerequisite of the caste system is that the members of a caste should have more or less the same degree of occupational prestige but not necessarily the same type of occupation. Such an assumption has important implications for the relationship between the caste system and the economy when one or the other changes.

If it is assumed that for the persistence of the caste system the members of a caste have to follow the same type of occupation, then with the transformation of the economy which gives rise to new kinds of occupations, there would also be a transformation of the caste system into some other form of social stratification. If it is assumed further that in the caste system the prestige of occupations rather than their identity is important, then the caste system can survive the economic change, provided the members of a caste follow occupations of an equal degree of prestige, even though the types of occupations followed are different.

The purpose of this paper is to examine the changes brought about in the caste system on account of marked changes in the economic organization of a community. The community chosen is Rayya, a small town in Punjab located along the Grand Trunk Road at a distance of 20 km from Amritsar in the direction of Jalandhar. For as long as its occupational composition is known, Rayya has been developing as a non-agricultural settlement because it is located en route for people visiting Baba Bakala, a Sikh pilgrim centre. During 1971–81 it emerged as a trading town and was one of the 34 settlements classified as towns for the first time in the 1981 Census.

The data used in this paper were collected in 1981 from a systematic random sample of 100 households, from a total of 1,602 households existing at that time. Two variables are specially important for the analysis: caste status and occupational prestige. Various studies in this region indicate that the relative status of a caste group may vary from place to place depending upon the variation in its socio-economic background. However, the relative prestige of an occupation is more stable. Therefore, the status of the various castes in Rayya was ascertained by asking a sub-sample of 25 respondents to rank-order the castes. However, for measuring the prestige of occupations, an occupational prestige ranking index with its seven-fold broader classification, prepared by the Population Research Centre of the Department of Sociology, Punjab University, was utilized.

Rayya had experienced rapid change both as regards population growth and occupational structure during the past two decades. Its population rose from 3,214 in 1961 to 4,556 and 7,049 in 1971 and 1981 respectively, representing growth rates of 41 and 54 per cent during the 1961–71 and 1971–81 decades, which are much higher than the corresponding growth rates for the state as a whole.

Rayya's rapid growth was due naturally to a relatively large number of immigrants.In the sample of 100 household heads, as many as 69 had migrated to the place during their lifetime and only 31 were born there. Among the immigrants 29 per cent came during 1961–71 and 52 per cent during 1971–81, which were the decades of rapid population growth.

Table 8.1, giving the percentage distribution of workers in the various industrial categories from 1961 to 1981, shows a remarkable sectoral transformation in Rayya since 1961. The main feature of the transformation is that the economy, which was predominantly non-agricultural even in 1961, had become overwhelmingly commercial in 1981. Trade has given the main impetus to immigration—as many as 80 per cent of the immigrants follow trading occupations, as against 48 per cent following similar occupations among the original residents. The transformation of the economy has resulted not only in inter-generational occupational mobility but also in intra-generational mobility. Among the household heads, 53 per cent had shifted from their first occupation; the corresponding percentage among the immigrants was naturally higher, which was 61.

One would be curious to know about the changes in the caste system due to migration and occupational mobility brought about by economic transformation. Our investigation reveals that Rayya's residents are able to identify the various castes in the community as well as their own in an unambiguous manner. The sub-sample of 25 respondents have also been able to rank the castes in a status hierarchy with a reasonable degree of consensus, as may be seen in Table 8.2. The

Table 8.1 *Percentage Distribution of Workers in Rayya by Industrial Categories (1961–81)*

Year	*Industrial Categories*									
	Cultivators	*Agricultural Labourers*	*Forestry, Mining, etc.*	*Household Industry*	*Manufacturing*	*Construction Work*	*Trade*	*Transport*	*Service*	*Total*
1961	22.87	5.74	1.08	13.92	4.69	4.67	17.72	5.72	23.59	100
1971	19.57	20.64	0.32	5.90	6.03	2.44	22.43	3.13	19.50	100
1981	7.00	10.00	–	6.00	3.00	–	60.00	–	14.00	100

Source: Information for 1961 and 1971 is taken from the Census reports and refers to the total universe. Figures for 1981 are taken from the sample survey.

Table 8.2 *Ranking of Castes in Rayya*

Rank in Descending Order of Status	*Caste*	*Median Ranking Score*	*Quartile Difference*
1	Brahmin	0.86	0.98
2	Jat	1.23	0.99
3	Khatri	2.40	0.80
4	Arora	3.40	0.92
5	Bania	4.30	0.63
6	Suniar	5.90	1.16
7	Chhimba	6.90	1.95
8	Mehra	7.00	2.36
9	Ramgarhia	9.00	2.25
10	Lohar	9.20	2.63
11	Nai	10.10	2.00
12	Dhobi	10.20	2.56
13	Kumhar	12.80	0.93
14	Ramdasia	13.20	0.94
15	Mazhbi Sikh	14.50	0.50

degree of consensus, as indicated by the quartile difference, however, is not uniform in all cases; generally it is less in the case of castes in the middle of the range (a finding which agrees with the results of most ranking studies). It is also evident from the median ranking scores that the respondents are unable to make much distinction in status between Chhimba and Mehra, Ramgarhia and Lohar, and Nai and Dhobi. On the whole, however, it turns out that the caste system is a viable form of social stratification in Rayya, despite the fact that the size of the sample used in our study of caste ranking is rather small.

As already pointed out, the status of castes is determined by the prestige of their members' occupations. Therefore, the assumption about the functioning of the caste system in Rayya has to be justified on the basis of the existence of a differentiation in the prestige of the occupations of different castes. Table 8.3 shows the distribution of the heads of households according to their caste and occupation. Since the castes are arranged in descending order according to status, and the occupations in descending order according to prestige, the table shows the relationship between caste status and occupational prestige. The average occupational prestige scores of the castes are shown in the last row. It is clear that, by and large, the higher castes follow occupations of higher prestige, and the lower ones occupations of

Table 8.3 *Distribution of Respondents by Caste and Occupation*

Occupation / *Caste*	*Principal* (6)	*Wholesaler* (6)	*RMP/Doctor* (5)	*Flour Miller* (5)	*Medium Shopkeeper* (4)	*School Teacher* (4)	*Medium Farmer* (4)	*Carpenter* (3)	*Petty Shopkeeper* (3)	*Small Farmer* (3)	*Clerk* (3)	*Driver* (3)	*Skilled Labourer* (2)	*Police Constable* (2)	*Tola (Carrier)* (2)	*Washerman, Barber* (1)	*Cobbler* (1)	*Watchman, Peon* (1)	*Domestic/Shop Servant* (1)	*Sweeper* (1)	*Agricultural Labourer* (1)	*Total*	*Av. Occ. Prestige*
Brahmin		2	2		3					1												8	4.6
Jat		2			1		2			2												7	4.2
Khatri		7		1	10	1			3													22	4.5
Arora	1	3			5				2	1												12	4.4
Bania		1							1													2	4.5
Suniar					3					1												4	3.7
Chhimba									2													2	3.0
Mehra									5			1							1			7	2.5
Ramgarhia		1			1			1														3	4.3
Nai									2		1			1					1			5	2.4
Dhobi									1							1						2	2.0
Kumhar					2																	2	4.0
Ramdasia					1				1						2		2		2	3	2	13	1.5
Mazhbi Sikh													1		2		1	1		4	2	11	1.2
Total	1	16	2	1	26	1	2	1	17	5	1	1	1	1	4	1	3	1	4	7	4	100	3.4

Note: Castes have been arranged according to the descending order of their status, and occupations according to the descending order of their prestige. Figures in brackets are the scores of occupational prestige.

lower prestige. There are, however, two notable exceptions—the average occupational prestige scores of Ramgarhias and Kumhars are much higher as compared with their caste status. It is obvious that the members of these two castes have achieved higher occupational prestige but their position in the caste hierarchy is based on the prestige of their traditional occupations of carpentry and pottery respectively. The discrepancy may be due to various reasons, such as an inadequate number of members of these castes in the sample and their predominantly immigrant status in the community (all the three Ramgarhias and one of the two Kumhars are immigrants). The Jats too have relatively lower occupational prestige, but they occupy the highest rank in ownership of wealth, as will be shown presently. Such discrepancies apart, there seems to be a significant degree of relationship between caste status and occupational prestige (the value of the rank order coefficient of correlation is –0.822).

Caste status is also related to certain other important aspects of the socio-economic status of the families, such as the median value of their property and the average score on the index of level of living of the members of a caste (see Table 8.4). The index of level of living was constructed on the basis of ownership of certain household effects, such as type of house, number of rooms, radio, television, certain items of furniture, and so on. The possible index scores range from 0 to 11.

Table 8.4 ***Median Value of Property Owned and Average Score on the Index of Level of Living***

Caste	*Median Value (Rs)*	*Rank*	*Average Score on Index of Level of Living*	*Rank*
Brahmin	62,500	3	8.7	1
Jat	2,25,000	1	8.4	3
Khatri	40,000	5	7.8	4
Arora	60,000	4	8.6	2
Bania	30,000	6	7.5	5
Suniar	20,000	8	6.0	8
Chhimba	10,000	9	4.5	10
Mehra	6,700	10	3.7	11
Ramgarhia	75,000	2	7.5	5
Nai	5,000	11	4.8	9
Dhobi	0	12	2.5	13
Kumhar	30,000	6	7.5	5
Ramdasia	0	12	3.3	12
Mazhbi Sikh	0	12	1.3	14

The ranks of the various castes in property ownership as well as in the scores on the index of level of living are more or less consistent with their ranks in caste status (the value of the rank order coefficient of correlation between caste status and property ownership is 0.756 and between caste status and level of living is 0.782). The major inconsistencies in these relationships exist in the case of Ramgarhias and Kumhars whose occupational prestige is also much at variance with their caste status.

On the whole, therefore, the positions of the various groups in the caste hierarchy agree with their differential socio-economic status. The Ramgarhias and Kumhars, though not present in sufficient numbers in Rayya, have moved appreciably higher in occupational prestige compared to their traditional status. This higher occupational prestige, however, has yet to be translated into higher caste status.

The given state of affairs raises the question: how has the stability of the caste system been maintained despite the far reaching changes in the economy? The stability of the caste system seems to have been accomplished through two main processes. First, by inducing migration mostly of those persons whose traditional occupations are akin to the new occupations generated by the changing economy, and second, by ensuring that the new opportunities go to people in accordance with their position in the caste structure, the higher castes benefiting more than the lower.

The operation of the forementioned processes in Rayya can be gleaned from Table 8.5, which shows the percentage distribution of heads of households and their average occupational prestige according to caste among the original residents and immigrants. The three trading castes of Khatri, Arora and Bania are grossly over-represented among the immigrants as compared with the original residents, their combined percentage being 46.4 and 12.9 respectively.

The operation of the second process is obvious from the fact that in almost every caste group in Table 8.5, there is a correspondence between the average occupational prestige scores of the original residents and the immigrants. The immigrants belonging to the higher castes, on the whole, have managed to secure occupations of higher prestige, and those belonging to the lower castes have secured mainly occupations of lower prestige. This impression is supported by the evidence about reasons for adopting the present occupation, which were obtained from 53 heads of households who had changed their occupation. Table 8.6 illustrates the reasons according to the respondents' castes, which are arranged in descending order according to

Table 8.5 ***Percentage Distribution of Heads of Households and their Average Occupational Prestige According to Caste among Original Residents and Immigrants***

Caste	*Original Residents*		*Immigrants*		*Total*	
	Households (%)	*Av. Occ. Prestige*	*Households (%)*	*Av. Occ. Prestige*	*Households (%)*	*Av. Occ. Prestige*
Brahmin	6.4	4.5	8.7	4.6	8	4.6
Jat	9.7	3.6	5.8	4.7	7	4.2
Khatri	9.7	4.0	27.6	4.6	22	4.5
Arora	3.2	6.0	15.9	4.2	12	4.4
Bania	–	–	2.9	4.5	2	4.5
Suniar	12.9	3.7	–	–	4	3.7
Chhimba	3.2	3.0	1.4	3.0	2	3.0
Mehra	19.5	2.6	1.4	3.0	7	2.5
Ramgarhia	–	–	4.2	2.0	3	2.0
Nai	6.4	3.0	4.4	2.0	5	2.4
Dhobi	–	–	2.9	2.0	2	2.0
Kumhar	3.2	4.0	1.4	4.0	2	4.0
Ramdasia	9.7	1.0	14.5	1.7	13	1.5
Mazhbi Sikh	16.1	1.4	8.7	1.3	11	1.2
Total	100.0		100.0		100	

Table 8.6 ***Respondents who have Changed their Occupation, by Caste and Reason for Adopting the Present Occupation***

Caste	*Reason for Adopting Present Occupation*				
	Better Income	*Only Available Employment*	*Less Investment*	*Miscellaneous*	*Total*
Brahmin	1				1
Jat	2			1	3
Khatri	15			1	16
Arora	6		1	1	8
Bania	1				1
Suniar					0
Chhimba					0
Mehra	3		1		4
Ramgarhia	1				1
Nai	1	1	2		4
Dhobi	1		1		2
Kumhar					0
Ramdasia		4	2		6
Mazhbi Sikh	1	4	2		7
Total	32	9	9	3	53

their status. Invariably, members of higher castes have given 'better income' as the reason while those of lower castes have mostly stated either that it was the only employment available or that the occupation required a lower amount of investment. Lucrative occupations in trade obviously involve substantial investments which only persons belonging to higher castes can afford.

The caste system has thus played a role in influencing the prestige of the occupations of persons even when they might have secured new types of occupations, inasmuch as persons of lower caste holding occupations of lower prestige in the past are still holding occupations of lower prestige in the new economy, although the jobs they are now doing are different from the previous ones.

There are instances of persons whose jobs in the new economy have much higher prestige than their earlier jobs. The influence of the caste system can, however, still be seen in the type of new jobs they have acquired. Attention may be focused only on persons belonging to the non-trading castes who are engaged in such trading occupations as wholesalers, medium shopkeepers and petty shopkeepers. In the region in which Rayya is located, generally the members of such castes as

Khatri, Arora, Bania and, to some extent, Brahmin are engaged in trade. For the other castes, trade is a new occupation. Therefore, the pursuit of commercial occupations by castes such as Jat (cultivator), Suniar (goldsmith), Chhimba (tailor), Mehra (domestic servant), Ramgarhia (carpenter), Nai (barber), Dhobi (washerman) and Ramdasia (cobbler), as shown in Table 8.1, deserves special notice.

A detailed examination of the trading activities followed by members of the forementioned non-trading castes (not shown in the table) reveal that the different castes, by and large, deal in different lines of trade, and the line of trade pursued by the members of a caste has, in most cases, some important link with their traditional occupation. For example, one of the Jats is a wholesaler in foodgrains, a job he was accustomed to handling in his previous occupation as a cultivator. The Suniars, formerly goldsmiths, have opened jewellery shops. The Chhimbas, traditionally tailors, deal in textiles. Some of the Mehras, traditionally domestic servants and therefore well versed in cooking, have opened restaurants. Among those Ramgarhias who have deviated from their traditional occupation of carpentry, one is flourishing as a saw mill owner and the other as a timber merchant, both handling the same kind of material as carpenters do. The Ramdasias, formerly cobblers, have shops dealing in leather goods, including shoes. And the Kumhars, traditional potters, are now selling aluminium and stainless steel utensils instead of earthen ones. Thus, in most cases the trading occupations newly acquired by the members of the non-trading castes have important links with their traditional caste occupations. In some cases, however, the new occupations (such as those followed by the Ramgarhias and Kumhars) have much higher prestige compared to their traditional occupations. These members of backward castes have been spared the competition from members of higher castes in their lucrative occupations because of the caste-linked skills necessary for their successful pursuit.

The discussion so far makes it clear that the existing social structure plays an important role in channelling changes in the economy, particularly in the occupational structure. In most cases, when an individual secures a new kind of occupation, he is restricted to occupations whose prestige is consistent with the status of his caste. On the other hand, when an individual has secured an occupation of much higher prestige compared with his caste status, his new occupation is in some way linked with his caste-related occupation. In either case, the caste background of a person has an important bearing on the opportunities available to him in the new economy.

It is important to underline the fact that sometimes persons belonging to lower castes too are able to achieve occupations of higher prestige normally pursued by higher castes. But it is also noteworthy that it is the particular caste background of these persons, such as that of the Ramgarhia and Kumhar, which has given them an edge over persons belonging to the higher castes in securing those occupations. Similarly, it is generally found in Punjab, where Rayya is located, that persons belonging to the blacksmith and cobbler castes have succeeded better in small-scale industries concerning metals and leather respectively. It is as if such persons have been provided with social immunity by the caste system, against competition for better economic opportunities from persons belonging to the higher castes.

The caste system has hitherto remained stable despite far-reaching changes in the economy, because of the abiding influence of the social structure in determining the economic opportunities of persons. Some castes, however, have succeeded in raising their relative position in the occupational-prestige hierarchy of castes although, for various reasons, their improved occupational status has not yet been translated into enhanced caste status.

The stability of the caste system in Rayya, despite economic changes, becomes further evident when we compare the average occupational prestige of the various castes over four generations: grandfathers, fathers, respondents and sons. The relevant information is presented in Table 8.7 and the rank order coefficients of correlation between different pairs of generations are given in Table 8.8. There is a remarkable degree of similarity in the relative positions of the various castes in occupational prestige between any two generations. Consideration of the respondents' sons' generation is rather premature as some of the sons have yet to join the labour force, and those who have joined have still to fulfil their employment potential. All the same, the correlation between the generations of the respondents and their sons is also significantly high.

The influence of social structure upon the economic opportunities of the members of the community naturally inhibits intergenerational mobility as regards occupational prestige, even though many persons might have changed their occupation from one type to another. An examination of intergenerational occupational mobility in Rayya also points to a meagre amount of occupational-prestige mobility taking place in the community over successive pairs of generations, as shown in Table 8.9. However, over the generations there is a trend towards

Table 8.7 ***Profile of Average Occupational Prestige of Various Castes in Four Successive Generations***

Caste	Generations			
	Grandfathers	*Fathers*	*Respondents*	*Sons*
Brahmin	4.6	4.6	4.6	6.2
Jat	3.4	3.4	4.2	4.8
Khatri	3.9	3.5	4.5	5.1
Arora	3.1	4.0	4.4	5.5
Bania	3.5	3.5	4.5	5.5
Suniar	2.2	3.2	3.7	5.6
Chhimba	3.5	3.5	3.0	3.0
Mehra	1.0	1.8	2.5	3.0
Ramgarhia	4.3	3.3	4.3	4.3
Nai	1.0	1.4	2.4	3.2
Dhobi	1.0	1.0	2.0	4.0
Kumhar	3.0	3.5	4.0	4.0
Ramdasia	1.0	1.0	1.5	3.2
Mazhbi Sikh	1.0	1.0	1.2	2.8

Table 8.8 ***Rank Order Coefficients of Correlation of Average Occupational Prestige of Various Castes among Different Generations of Respondents***

Pairs of Generations	*Coefficients of Correlation*
Respondents' fathers and grandfathers	0.787
Respondents and their fathers	0.838
Respondents and their sons	0.800
Respondents and their grandfathers	0.824

Table 8.9 ***Pearson's Product Moment Coefficients of Correlation of Intergenerational Occupational Mobility for Different Pairs of Generations***

Pairs of Generations	*Coefficients of Correlation*
Respondents' grandfathers and fathers	0.875
Respondents and their fathers	0.791
Respondents and their sons	0.619
Respondents and their grandfathers	0.741

increasing mobility. Mobility from the respondents' to their sons' generation is relatively high, but this is partly because of the special reasons pointed out in the foregoing.

In evaluating the change and continuity in social structure, one has to keep in view two important considerations: first, there has been a far-reaching change in the economy of the community and, second, the size of the sample studied is rather small. Both these considerations would warrant one's expecting to find a larger degree of change in the caste structure, but what one finds is a relatively large degree of stability. Therefore, one may conclude that despite fundamental changes in its economy the social structure of Rayya has continued to be relatively stable.

Rayya is typical of a large number of towns in Punjab growing in commercial activity as a result of the green revolution (D'Souza 1976). It would not be unreasonable, therefore, to generalize the results of Rayya for Punjab as a whole, and to suggest that the rapid economic changes that have been taking place in Punjab during the last three decades or so have not resulted in a transformation of the social structure of the region.

REFERENCES

D'Souza, Victor S. 1976. 'Green Revolution and Urbanization in Punjab during 1961–71'. *In* S. Manzoor Alam and V. V. Pokshishevsky (Eds.), *Urbanization in Developing Countries*. Hyderabad: Osmania University.

9

The *Ardhatiyas* in a Small Town in Uttar Pradesh: A Study in the Sociology of Business

KHADIJA A. GUPTA

INTRODUCTION

My acquaintance with the traders called Ardhatiyas (commission agents, middlemen) dates back to the sociological study of 'small town politics' I conducted in Ranipur, Aligarh district, Uttar Pradesh, during 1968–71 (see Gupta 1975). Though the study did not focus on the Ardhatiyas *per se*, I was surprised to find that studies on the Ardhatiyas were not only very few but nearly all of them dealt with issues of concern to economists. Studies so focused are no doubt valuable. However, I was not looking for these. I wanted studies dealing with small town Ardhatiyas from the sociological perspective. I could not locate even a single study.

It was this gap which prompted me to undertake this study.[1] My study was conducted in Sikandra Rao, a small town in Aligarh district. It is an exploratory study seeking to understand, primarily through field investigation, the social and economic roles of the Ardhatiyas as reflected in their network of relations, activities and methods of working in a small town setting. In the absence of systematic studies defining the boundaries of a group like the Ardhatiyas, I felt that the best course would be to use the Ardhatiyas' own frame of reference as a basis for understanding their social role and organization. The initial data were collected through formal and informal interviews, discussions and participant observation. These were subsequently supplemented by more formal techniques, such as open-ended questionnaires and opinion surveys.

Sikandra Rao is the headquarters of one of the six tehsils of Aligarh district. The tehsil, named after the town, consists of 235 villages. Among the five towns in the district, Sikandra Rao with a population of 17,000 (1971 Census), covering an area of 0.65 sq km, ranks third. The four other towns are Aligarh (population 252,300), Hathras (74,300), Sasni (6,900) and Mursan (5,300). Sikandra Rao is located 39 km south-east of Aligarh, the headquarters of the district.

THE *ARDHAT* TRADE: A HISTORICAL REVIEW

The word *ardhat* occurs in varying forms in practically every Indian language: *ardhat* in Hindi and Maithili; *arat* in Bengali and Sindhi; and *adat* in Gujarati and Marathi. Hence the derivatives *ardhatiya* (Hindi), *aratdar* (Bengali), *adatiya* (Gujarati, Marathi), *adathi* (Tamil) and *arti* (Punjabi). The word was well in use in commercial parlance during Buchanan's survey of the Bengal and Bihar districts during the early nineteenth century (cited in Ray 1982). He gives instances of the various interrelated uses of the term in his account of Patna. He also mentions bankers known as Aratiyas in smaller towns who granted *hundis* for cash in their capacity as correspondents to bigger banking and trading houses in the major marketing centres. He distinguished

[1] This paper is based on a research project entitled 'Small Town Ardhatiyas: A Sociological Study of their Role, Organization and Entrepreneurial Charade' sponsored by the Indian Council of Social Science Research, New Delhi.

the Ardhatiya's three roles: (*a*) acting as commission agents for sale and purchase on behalf of others; (*b*) maintaining warehouses for stocking and commissioning the sale of grain; and (*c*) acting as correspondents of bankers on which they could issue *hundis*.

According to Ray, the indigenous commission agency system not only successfully resisted the intrusion of Western methods of business and Western types of organization but also took the place of wholesale trading. He further says:

> Wholesale firms were almost invariably commission agency firms, which bought and sold not so much on their own account as on behalf of other buyers and sellers, whom they financed to the extent of three-quarters or more of the goods and produce which they handled on their behalf. Under the commission agency system, losses in transit would never fall on the commission agent, but would be borne by the consignee. On the other hand, bad debts would fall on the commission agent (Ray 1982: 74).

The system owed its prevalence also to the crucial position it occupied in the credit structure of the country. Credit had to be given, otherwise trade would stop. The commission agent handling goods on behalf of borrowers used as collateral for his lendings the goods which he bought and sold for his indebted clients. The poorly developed farms across the country needed advances for production. The safest form of credit was an advance against a crop or against goods in transit.

In *mandi* (market) towns, the Ardhatiyas figured mainly as commission agents through whom the peddler or the village producer sold his crops or, conversely, as commission agents through whom other manufacturing centres sold their goods to retailers. In general, the cultivator's crop passed through four middlemen: the village shopkeeper (a peddler variously known as *beopari*, *faria* or *dalal*) who brought the produce to the *mandi* and sold it through a commission agent, who usually took the assistance of a broker, to a *mandi* merchant.

There were two kinds of commission agents: *kuccha* Ardhatiyas and *pucca* Ardhatiyas. According to Ray:

> The *kuccha* Ardhatiya was basically an intermediary, whose function was to introduce the *beopari* to a purchaser, and to arrange a bargain usually with the help of the purchaser's *dalal* or directly

> with the purchaser. He did not buy on his own account; he was merely the commission agent of the village dealer, and his aim was to secure the custom of as many village dealers as possible by making them advances for the produce. The *pucca* Ardhatiya, on the other hand, was a true commission agent, who bought on behalf of some wholesale firm in another city. Thus the purchaser to whom the *kuccha* Ardhatiya introduced the *beopari* was usually a *pucca* Ardhatiya (1982: 70).

The difference in the role of the two Ardhatiyas thus lay in the interests that differentiated their respective clients, with the *kuccha* Ardhatiya seeking to make as good a bargain as possible for his village customer by selling dear at one end, and the *pucca* Ardhatiya trying to purchase as cheaply as possible on behalf of his wholesaler-customer, usually in another *mandi* at the other end. In the grain business the *pucca* Ardhatiya bought on behalf of other merchants as well as on his own account, but in the business of other products he never dealt on his own account. In the bigger *mandis* the *pucca* Ardhatiya predominated. The *kuccha* Ardhatiya generally confined his business to the smaller and more dispersed *mandis* of the poorer agricultural trades. He was a mere collecting agent for the countryside, passing the produce on to the Ardhatiya in the larger *mandi*, who often financed him. Usually the *pucca* Ardhatiya did not buy directly from the peasants or the *beoparis*; he dealt with them through the *kuccha* Ardhatiya. It was to him, rather than to the *kuccha* Ardhatiya, that the miller or the wholesaler in another centre sent his agent. The *pucca* Ardhatiya in turn sold to the retail traders. He was thus the primary distributor of agricultural produce.

The Ardhatiyas employed, as they do now, several market personnel—*dalals* (broker), *tauldars* (weighmen), *hammals* and *palledars* (labourers who attended to cleaning and handling produce), *munims* (clerical staff), *shagirds* (apprentice partners), sweepers, water carriers, and so on. Besides his own commission (*ardhat*), therefore, the Ardhatiyas charged for other services also, namely, *dalali* (brokerage), *palledari* (handling charge), *tulai* (weighing charge), *karda* (quality deduction), *dhalta* (weight deduction), and miscellaneous, particularly *dharmada* (donation for religious purposes) and *goshala* (food and shelter for cows and other animals in distress). About the last, Ray writes:

> The rates for *dharmada* and *goshala* were fixed by the Ardhatiyas' market panchayat, but the income was spent at the discretion of the

> individual Ardhatiya, who was not accountable to anyone. Some Ardhatiyas devoted the amount to schools, orphanages and other charitable institutions but others—the majority—spent it on religious ceremonies, feasting of Brahmins, pilgrimages to Hardwar and Benaras, bathing in the sacred waters of the Ganges, the upkeep of temples, and the maintenance of alm houses for old and infirm cattle (1982: 73).

The dealings between the Ardhatiyas and the peasants were, to a large extent, direct. The Ardhatiya generally acted in the interest of the cultivator as the volume of his business with the peasant sellers depended on his reputation. The same group of peasants would employ him year after year, showing their confidence in him, though of course there were individual cases of complaint that the Ardhatiya was cheating the cultivator.

Quite a few local castes participated in the peddling trade as *farias*, *beoparis* and *paikars*, operating in a single rural locality or moving through the country as itinerant traders. They were the linkmen of the 'market' and had a foothold in the market town. But the Banias dominated the trade that functioned through bills of exchange and commission agencies. Regionally, the Banias included at least three distinct groups: (*a*) Agarwals and other Hindi-speaking Vaish castes of the upper Gangetic valley, Bundelkhand and Delhi; (*b*) Hindu and Jain Marwaris hailing from the former princely states of Rajputana, mainly Bikaner and Marwar, and from the Sekhawati region of northern Rajasthan and southern Haryana; and (*c*) the Gujarati Banias, both Hindu (Meshri) and Jain (Shravak).

> The Bania communities formed the principal element in the commercial population of Rajputana and Gujarat, but west and south of these areas, there were other Hindu merchant communities—the Khatris and Aroras of the Punjab, the Lohanas of Sind (among whom figured the Multani bankers of Bombay), the Bhatias of Kutch, and the Chettiars and Komatis of the Madras Presidency. Bombay city was, moreover, the base of the Parsees and of three great Muslim merchant communities converted from amongst Hindu merchants—the Bohras, the Khojas and the Memons (the last two communities being Muslim converts from the Lohanas of Sind). It should be noted that these large merchant communities were divided into numerous smaller endogamous communities,

> some of which, such as the Maheshwaris of Bikaner, the Oswals of Marwar, the Agarwals of the upper Gangetic valley, the Kapol Banias of Bombay, the Shikarpur Shroffs (Lohanas known in Bombay as Multanis), the Cutchi Memons, and the Nattukottai Chettiars, distinguished themselves by their enterprise (Ray 1982: 75).

The Bania communities controlling the market enjoyed higher status than that of the mass of retailers and peddlers. Often, the latter belonged to altogether different ethnic or linguistic groups. This was not so evident in the homeland of the Hindu and Jain Bania communities, however. The Marwaris of Rajasthan, the Meshris and Shravaks of Gujarat, and the Agarwal and other Vaish communities of Uttar Pradesh and Haryana were numerous in their native places and were thus to be found at all levels of trade and finance—from the village *bania* to the all-India financier. But away from the heartland of these communities, caste and ethnic distinctions tended to reinforce the distinction between the top controlling level of the 'market' and the peddling and retailing levels of rural and urban trade.

SIKANDRA RAO AND ITS TRADERS

The pattern of *ardhat* trade outlined in the previous section does not seem to have undergone major changes. Some changes have no doubt taken place, particularly during the post-independence period when the government mounted massive efforts to promote development in various walks of life and also made legislative and other interventions to regulate the trade. Changes have also taken place in the functioning and role of the Ardhatiyas. However, these have not radically altered the character of the trade. Continuity amidst change seems to be the dominant theme of the development process under way. I shall examine this theme by describing some aspects of the town of Sikandra Rao and of the Ardhatiyas and their trade in the town.

Very little is known about the growth of Sikandra Rao as a marketing centre or *mandi*. It is clear, however, that the town has been in existence for several centuries. It is said to have been founded by Sultan Sikandar Lodhi in the sixteenth century. According to the *District Gazetteer* of 1926, it was known as Sikanderpur. Subsequently it was given in *jagir* to an Afghan named Rao Khan. Since then it has been known by its present name Sikandra Rao. It had been a

Pathan settlement and the seat of Muslim governors for long. Until the Sepoy Mutiny of 1857 the Pathans settled in the town wielded considerable power and owned large estates. They lost their fortunes during the Mutiny when the Rajputs in the neighbourhood took over the town. Later, one of the Rajput leaders, Thakur Kundan Singh, who opposed the Pathans and supported the British, was rewarded for his services. He was made the tehsildar of the *pargana*. Another person who received favours from the British rulers for his services was Debidas, 'a prominent Bania of the place' (Nevill 1926: 290).

The population of the town has steadily increased from 10,950 in 1951 to 13,899 in 1961 and 17,253 in 1971. At one time the Muslims constituted the majority. The Hindus now occupy this position, though the Muslims still constitute a substantial segment of the town population. The census records show that the proportion of Muslims declined from 49.4 per cent in 1901 to 41.7 per cent in 1971, and that of Hindus (including Scheduled Castes) increased during the same period from 48.0 per cent to 57.4 per cent. Excluding the Scheduled Caste population, the Hindus constituted 44.6 per cent of the town population in 1971 (see Table 9.1).

Table 9.1 ***Distribution of the Population of Sikandra Rao by Religion, 1971***

Religion	*Population*		*Percentage*	
Hindu	9,912		57.4	
Upper castes		7,700		44.6
Scheduled Castes		2,212		12.8
Jain	114		0.7	
Muslim	7,202		41.7	
Other (Sikh, Christian)	25		0.2	
Total	17,253		100.0	

According to the *District Gazetteer* of 1926, the principal Hindu residents of the town were Banias, a few of whom owned large estates in the neighbourhood. Most of them, however, followed their traditional occupation—trading, commission sales, brokerage and moneylending. The Banias still dominate the town's economy. Almost the entire wholesale trade in grains is in their hands. Nearly all Ardhatiyas, *kuccha* or *pucca*, small or big, belong to this caste, consisting mainly of two subcastes, Maheshwary and Varasheny. Some of them also operate as peddlers and retailers. Most of the rice and flour mills and other manufacturing units (like oil crushing) are owned and managed by members of this caste.

Trading and commercial activities in Sikandra Rao are primarily confined to the marketing of agricultural produce, mainly foodgrains. According to the Uttar Pradesh Agricultural Marketing (Regulation) Act of 1964, Sikandra Rao is a B Class *mandi* (grain marketing centre).[2]

The Agricultural Produce Marketing Committee (APMC) of the town constituted under the Agricultural Marketing Act is responsible for regulating the marketing of agricultural produce (worth about Rs 30 million per year). The major crops grown in the area and marketed by the town Ardhatiyas are wheat, gram, mustard, oilseed, pulses and peas (*rabi* crops), and paddy, maize, *jowar* and *bajra* (*kharif* crops).

Local accounts suggest that till 1955 the town looked like a nodal rural market centre of the kind one sees in large and well populated villages in the district. Only a small quantity of foodgrains used to be marketed by half-a-dozen *kuccha* Ardhatiyas, who combined the activity with moneylending and retailing. All these Ardhatiyas were of village origin, some hailing from far away villages in the Marwar region of Rajasthan.

The spurt in agricultural production in the region began, as in many other parts of the country, in the mid-fifties following the launching of a spate of development programmes, including programmes to build roads and other infrastructure in the countryside. That these developments helped the growth of the town is reflected in local accounts as well as in the rise in the town's population by as much as 25.5 per cent during 1951–61. The following decade also recorded a high growth (about 24 per cent), though a little less than in the previous decade. The number of Ardhatiyas and *ardhat* firms also increased from 6 in the 1950s to over 20 in 1980. At the time of my enquiry (1981), the town had 25 registered *ardhat* firms. There was no change in this number in 1983 when my enquiry concluded, but the trade per firm had expanded.

[2] The Act classifies *mandis* according to the range and volume of trading operations into four classes: Special A, A, B and C. Including Sikandra Rao, there are eight *mandis* in the district. Four of them are bigger, enjoying higher ranking than Sikandra Rao—Aligarh and Hathras (Special A Class) and Charra and Khair (A Class). The remaining three—Sasni, Atroli and Jattari—are C Class.

Most of the firms are family enterprises named after fathers or grandfathers who founded them, and some after the owners' minor sons. However, following the Agricultural Marketing Act (1964) and the constitution of the APMC (1967), there has been a shift in this practice. The trend is to register the firm as a partnership firm, even if it is owned by members of the same family. In some cases, even those who entered the trade several years before the Act was introduced and ran the firms as family enterprises, have re-registered them on partnership basis. The reasons for the emergence of this trend vary from case to case. Generally, it is due to the Ardhatiyas' perception or understanding of the benefits and limitations of the corporate and taxation laws and the measures introduced to regulate the trade. In some cases it is due to the desire on the part of the old founder of the firm to facilitate the transfer of the family legacy to his sons or other relatives without dividing it into separate units. In a few cases, it is due to two or more individuals, not necessarily bound by caste or kinship ties, wanting to enter the trade by pooling their resources and expertise.

Besides the registered Ardhatiyas, there are a number of unregistered intermediaries involved in the trade. There is the village moneylender-cum-trader who buys the produce from cultivators, transports it to the town, and sells it through some *kuccha* Ardhatiya. He buys the produce even before the crops are harvested, at times soon after they are sown. Otherwise he buys the produce immediately after the harvest. Not being a registered dealer, he tries to bring the produce to the town as a producer, and does it successfully without much difficulty. Not only is the mechanism to regulate the trade weak, but apparently it also offers scope for collusion between the 'regulator' and the 'regulated' to beat the system.

Another unregistered intermediary is the town trader who, in the opinion of some registered Ardhatiyas, operates as a 'floating' or *faria* Ardhatiya. He lives in the town, but collects the produce from the outskirts of the town where quite a few small cultivators send their wives and children to sell small quantities of grains in order to buy daily provisions like salt and oil. The 'floating' Ardhatiya waits for them with the merchandise and barters it to pick up the cultivators' produce. When his collection becomes sizeable, he sells it through one of the *kuccha* Ardhatiyas in the town. Another category of intermediary that has lately entered the field is the bullock cart transporter from the village. He collects and transports the produce from the farm to the town and sells it to one of the Ardhatiyas.

SOCIAL AND ECONOMIC BACKGROUND OF THE ARDHATIYAS

Of the 25 Ardhatiyas in Sikandra Rao, five are *pucca* and 20 *kuccha*. One of them, Radhey Shyam, is registered as both *pucca* and *kuccha*. Another, Kishan Kumar Tilak Raj, is registered as a *pucca* and *kuccha* Ardhatiya as well as a wholesaler. While all the 25 Ardhatiyas deal in foodgrains, only five were in this trade before 1955.

Virtually the entire *ardhat* trade in the town is in the hands of Banias, with 22 of the 25 Ardhatiyas belonging to this caste. They are divided into three subcastes—9 Maheshwarys, 9 Varashenys and 4 Mahajans and Mahours. The Maheshwarys and Varashenys enjoy a higher ritual ranking than the Mahajans and Mahours. It is not clear whether the Mahajans and Mahours represent a single subcaste, as the Maheshwarys and Varashenys in the town think, or whether they are two distinct subcastes, as the Mahajans and Mahours themselves think. For the purpose of this article, however, the issue is not as important as the fact that *ardhat* as well as other trades in the town are dominated by the Banias. The few processing and manufacturing units in the town are also owned largely by them.

Most Ardhatiyas in the town, especially those who have been in the trade for a longer period, had humble beginnings as village traders-cum-moneylenders. All of them entered the *ardhat* trade as *kuccha* Ardhatiyas. When they did this, quite a few also took to cloth trading as petty shopkeepers. They continue in this trade now. They seem to view the two trades as mutually reinforcing—one providing clientele for the other. Bhagwati Sahay, the founder of a leading *pucca ardhat* firm, said:

> Both cloth and grain trade attract a large clientele from the nearby villages. Most of them are small farmers who are always in need of credit. We sell them cloth on credit. If they also want to buy other commodities on credit, which they do, we send them to the traders concerned, guaranteeing payment for the goods sold on credit. Through the cloth shop we thus secure clientele for the *ardhat* firm. Over time we establish through such clients contacts

in their respective villages. They operate as our channels of information, and from amongst them we recruit our *munims* and *kamdars*. It is through them that we build our market network in the villages for the *ardhat* trade.

Six of the 25 firms in the town have been in existence for more than 50 years. The firms, named after the present owners, are:

Pucca ardhat: Radhey Shyam (Varasheny); Mangal Sen Ram Gopal (Mahour).

Kuccha ardhat: Nawal Kishore (Maheshwary); Bishamber Dayal (Maheshwary); Chiranji Lal Banwari Lal (Maheshwary); Kedar Nath Suraj Parshad (Varasheny).

The present owners' predecessors who founded the firms are no longer alive. But the families have kept alive the legacy, one generation bequeathing it to the next. Five of the firms are three generations old, while the sixth (Nawal Kishore) is four generations old. The families migrated to the town or to a village adjoining it several decades ago. They came from distant villages in Rajasthan. All are Banias, three of them Maheshwarys.

Socially, the Ardhatiyas in the town do not seem to enjoy high esteem. The people of Sikandra Rao in general, including businessmen of other categories and those belonging to the same caste as the Ardhatiyas, look down upon the Ardhatiyas, unlike their counterparts in Ranipur where they occupy high social status. The popular image projects them as a group of miserly people, uneducated and rustic, whose life revolves around *toul-bhao* (weighing and pricing), meaning petty profit making.

Whatever be the reason for the low status accorded to the Ardhatiyas by the town people, the Ardhatiyas themselves do not seem to mind it. Indeed, they seem to take pride in the view that their life revolves around *toul-bhao* and that they are not *babujis* (that is, rule-oriented, leisure-loving white-collar office-goers who take pride in their inability to do manual work). Thus, when I asked one of the *kuccha* Ardhatiyas about the duties and working hours of his staff and his own routine, he snapped back:

We do not like to be viewed as *babujis*. We do not hesitate doing such petty manual jobs as loading and weighing. We do not sit and wait for someone to do the work for us. We work as the work demands. To us the question of 'duties' or 'working hours' is irrelevant.

While to the town people the Ardhatiyas are an undifferentiated group, the traders in general and the Ardhatiyas in particular differentiate the ranks the individual Ardhatiyas occupy within the trade in terms of their social origin and *pucca-kuccha* classification. In general, the *pucca* Ardhatiyas are accorded higher status than the *kuccha* Ardhatiyas. This observation is based on the findings of an opinion survey I conducted. Thirty-five persons were randomly selected for this purpose from a wide range of traders dealing in agricultural produce: owners of agro-processing enterprises, petty traders and well-to-do traders, pavement squatters selling cheaper varieties of foodgrains and other commodities to poorer people of the town, and small-scale but economically well-off industrialists like rice mill owners. They were asked to rank each Ardhatiya in descending order. The ranking varied from one respondent to another. Not all respondents ranked all the Ardhatiyas, leaving out those who they felt were not worth ranking. Of the 25 Ardhatiyas, only five (four *pucca* and one *kuccha*) figure in the ranking list of each respondent (see Table 9.2).

Table 9.2 ***Ranking of the Top Five Ardhatiyas by the Town Traders***

Ardhatiya's Name	*Type of ardhat*	*Ranking*	
		Social	*Economic*
Jagdish Parshad	*Pucca*	I	II
Ram Gopal	*Pucca*	I	II
Radhey Shyam	*Pucca*	I	V
Kishan Kumar	*Pucca*	II	I
Suraj Parshad	*Kuccha*	II	I

That *pucca* Ardhatiyas occupy higher social and economic status than *kuccha* Ardhatiyas is accepted by both the groups of Ardhatiyas. One obvious reason is that, compared to *pucca ardhat*, *kuccha ardhat* is a low-risk, low-investment, low-turnover venture. Another reason is the difference in their roles and relationships. The *kuccha* Ardhatiya maintains a relationship with a limited group of producers and *pucca* Ardhatiyas. The latter, on the other hand, maintains a relationship with a wide range of *kuccha* Ardhatiyas and with peddlers, retailers, wholesalers, mills and processing units in large towns.

The social origin of the Ardhatiyas is another element in status differentiation. Take, for example, the following descriptions given by two well established *pucca* Ardhatiyas, one belonging to the family of a locally reputed zamindar and the other of humble origin but with greater success and prosperity in business. The first, Radhey Shyam, is the proprietor of Radhey Shyam and Co. Tracing his descent he said:

> We belong to a family of zamindars. We used to pay Rs 200,000 as land revenue. Our family owned land in 52 villages, 50 indigo pits, two glass factories and a number of houses in the town. When the ancestral property was divided, I got Rs 50,000 in cash as my share. During the Quit India movement, when I was a graduate student, I left my studies. On my return from jail, I established two glass factories with two Muslims as *munims* (accountants). During partition the accountants migrated to Pakistan, taking with them the entire cash that I kept in trust with them to run the factories. It was a great setback. The problem for me was how to support my family. I thought for days what I should do to start a new life. Eventually I decided to open a *kuccha ardhat* shop in the town. When I was young I had friends whose parents were in the *ardhat* business. I used to observe the way they operated the trade. Those memories were alive in my mind. Besides, being the descendent of an old zamindar family I had good contacts with the villagers in the neighbouring villages.
>
> I invested Rs 10,000 in the *kuccha ardhat* business. Initially business was slack, and supply was meagre. There were only two buyers—*pucca* Ardhatiyas—in the town. They had almost monopolistic control over the *kuccha* Ardhatiyas and they used to give a lot of trouble to them as well as to the village traders operating as unlicensed *kuccha* Ardhatiyas in the town. In 1955, I converted the *kuccha ardhat* into *pucca ardhat* mainly with a view to help the *kuccha* Ardhatiyas. I also hoped that by getting into this business I would come to know the various malpractices prevalent in the *pucca ardhat* trade. I wanted to make the *kuccha* Ardhatiyas aware of such malpractices and also force the *pucca* Ardhatiyas to mend their ways. To tell you the truth, most of the Ardhatiyas are illiterate villagers. As soon as they are able to save Rs 500 they want to

> become Ardhatiyas by getting into the *kuccha ardhat* business to begin with. They do this even though they know nothing about the business. That is why there is such a high rate of failure or mortality of firms in this business.

Bhagwati Sahay is the head of a family which owns one *pucca ardhat* firm, one wholesale firm and two cloth shops. Sahay himself is illiterate, but two of his eight sons are Commerce post-graduates, the third an MBBS medical doctor, the fourth an MBBS student studying at the Gwalior Medical College, and the remaining four college and university students of science and technology. Irrespective of the specialization or level of education, they have either come back to the town or hope to do so after completing their education and then join the family business. Describing the origin and growth of the family business, Kishan Kumar, the third son and a Commerce post-graduate, said:

> My father migrated to the town from Danapur village of Buland Shahar district. Originally we were from Rajasthan but later settled down in Buland Shahar. My father was a village moneylender-cum-trader. In 1940, he and one of his brothers, Jagdish Kumar, visited the town at the invitation of my maternal aunt who was settled in Sikandra Rao. As she had no child, she adopted my uncle, Jagdish Parshad. Her husband, Hazarilal, and my father opened a cloth shop in the town. Through this business my father built up contacts in the neighbouring villages and later opened a *kuccha ardhat* shop in the town. Once he accumulated sufficient money and 'resources' (that is, contacts) he opened a *pucca ardhat* shop and also acquired a licence to operate as a grain stockist. Besides these, our family now owns three cloth shops. As in the past, we hope to keep on expanding the family business.

Economically as well as in terms of the variety and scale of business, the Sahay family is more prosperous than the Radhey Shyam family. The former's investment in business is reported to be over Rs 40 lakh, several times more than the latter's. Likewise, the annual turnover of the units owned by Sahay is several times more than Radhey Shyam's. Yet, because of their humble origin, the town people, including those in trade and business, consider the Sahay family lower in status than the Radhey Shyam family. The latter is considered part of the town's gentry because of the family's respectable social background.

THE ORGANIZATION OF ARDHAT TRADE

The *kuccha* Ardhatiya's main function is to sell the produce of his farmer clients to the *pucca* Ardhatiyas. Notionally, he is the agricultural producer's agent. But the fact that he sells the produce to a *pucca* Ardhatiya, and gets his commission at the rate of 1.5 per cent of sale from the latter, makes him dependent on the *pucca* Ardhatiya. Often he works in collusion with the latter, who may be one of his close relatives or a member of his caste with whom he has many social ties. In practice, therefore, the *kuccha* Ardhatiya operates more as a subsidiary, if not an agent, of the *pucca* Ardhatiya. This contradiction between his publicly projected role and the implicit role makes him suspect as far as the producers are concerned and a subordinate as far as the *pucca* Ardhatiyas are concerned. It is not surprising therefore that the *kuccha* Ardhatiya enjoys a lower status than the *pucca* Ardhatiya.

The *pucca* Ardhatiya plays multiple roles. He is the real commission agent for outside parties as well as for himself if he happens to be a wholesaler, which he is in the town. He is also an agent of the Regional Food Corporation (RFC), a stockist, a moneylender or financier, and an entrepreneur, in the sense that he often diversifies his investments if only to exploit new opportunities or to settle his sons and relatives in similar or allied trades, even in industries involving low risk and technology. This in turn gives rise to the interlocking of investments and interests and a complex set of relations, both social and economic, legal and political.

All Ardhatiyas, whether *kuccha* or *pucca*, are required to register themselves with the APMC, locally called the Mandi Samiti. It is also the licence issuing authority. The licence rate and related charges (see Table 9.3) vary depending on the nature of *ardhat* (*kuccha* or *pucca*) and crop categories, namely, *anaj* (cereals), *dal* (pulses), *tilhan* (oilseeds) and *gur* (jaggery).

The *pucca* Ardhatiyas have to register themselves with and obtain licences from not only the APMC but also the RFC. Besides security deposits and licence fees, the APMC charges a cess of 1 per cent on the value of business transacted by each *kuccha* Ardhatiya. Because of the seasonality of trade, the volume of business and hence the amount of cess collected varies from season to season, month to month, and even day to day.

It is estimated that the total volume of business transacted by the registered Ardhatiyas in the town is in the range of Rs 30 million a

Table 9.3 *Licence Rates and Related Charges*

Ardhat/Crop	*Charges*		
	Security Deposit	*Registration & Licence Fee*	*Yearly Licence Renewal Fee*
Kuccha Ardhat			
Anaj (cereal)	500	75	35
Dal (lentil)	500	75	35
Tilhan (oilseed)	750	75	35
Gur (jaggery)	1,000	100	100
Pucca Ardhat			
Anaj (cereal)	2,000	500	35
Dal (lentil)	1,000	500	35
Tilhan (oilseed)	750	500	35
Gur (jaggery)	1,000	100	100

Note: All figures are in Rs.

year. There are no records to show the break-up of this turnover by crops. However, according to a rough estimate provided by one of the Samiti officials, the turnover of the important crops marketed by the town Ardhatiyas in 1979–80 ranged as follows:

Wheat : 4,000 tonnes
Paddy : 3,000 tonnes
Pulses : 5,000 tonnes
Oilseeds : 1,000 tonnes

The village cultivator wanting to sell his produce in the town has to obtain an entry permit from the APMC. According to APMC rules, the producer is expected to bring the marketable produce himself. He must be present in the *mandi* at the time his produce is sold or auctioned by the *kuccha* Ardhatiya. He is expected to sign the sale voucher, which the *kuccha* Ardhatiya collects and deposits with the APMC. These prescriptions, like some others, are rarely followed. In most cases the producer operates through an undeclared proxy, a village trader, a 'floating' Ardhatiya, or at times the *kuccha* Ardhatiya himself.

The APMC has not yet constructed any market yard. Instead it has declared the old grain market and some areas adjoining it as the town's temporary market yard. At this place there exist a number of shops-cum-godowns, or *ardhats* as they are locally called. The agricultural produce for sale is brought here by the producer or his nominee in bullock carts and dumped in front of the shop owned or operated by his agent, the *kuccha* Ardhatiya. The produce is then graded and kept in separate piles on the roadside. In case the produce is graded by the producer before unloading, the Ardhatiya indicates the spot where it is to be unloaded. When this is not the case, unloading and grading are done by hired labourers under the Ardhatiya's supervision. The producer is free to hire labourers of his choice from the market. Alternatively, he can engage labourers under the Ardhatiya's employ. In either case, the producer is required to pay the wages.

Normally the work of unloading and grading is completed before mid-day. Around 1 p.m., when the work is over, the *kuccha* Ardhatiya sends one of his relatives to meet each of the five *pucca* Ardhatiyas in the town and requests all of them to visit his *ardhat* and bid for the stock on sale. This is not a mere courtesy which the *kuccha* Ardhatiya extends. It is a business norm as well as a social tradition that he is expected to follow both in spirit and letter. Even if he deviates marginally from this practice, say, by sending one of his employees instead of a relative, the *pucca* Ardhatiya may view it as a serious lapse and may not visit his *ardhat* at all. Other *pucca* Ardhatiyas may also similarly punish him.

If the tradition is correctly followed, which the *kuccha* Ardhatiyas always do, the *pucca* Ardhatiya in turn is expected to visit the *kuccha* Ardhatiya's shop and participate in the bidding. However, the *pucca* Ardhatiya being in a better position, enjoying as he does greater command over the market than the *kuccha* Ardhatiya, occasionally ignores the *kuccha* Ardhatiya's invitation, particularly if the latter happens to be a new entrant.

The bidding is done on the spot and in writing. The *kuccha* Ardhatiya gives a small piece of paper to each *pucca* Ardhatiya present at the bidding, in the presence of the Kamdar (APMC inspector) as well as the producer. The *pucca* Ardhatiya bidder randomly picks up a handful of grains from the stock under bidding, examines the sample, writes down his quotation on the slip of paper given to him, folds it and hands it over to the *kuccha* Ardhatiya. When all the bidders participating in the bidding have handed over the folded slips, the *kuccha* Ardhatiya opens the slips. He announces the highest bid and asks the producer whether it is acceptable to him. If it is, the *kuccha* Ardhatiya loudly declares the deal as settled. The successful *pucca* Ardhatiya bidder is asked to sign the sale voucher—a form prescribed by the APMC—in triplicate. The *kuccha* Ardhatiya retains one of the copies, sends one to the *pucca* Ardhatiya who won the bidding and submits the third to the APMC.

While all this sounds fair and above board, there is a feeling widely shared by the APMC officials that the *pucca* Ardhatiyas of the town, being limited in number, often collude to keep the bidding low. This they do by agreeing beforehand who among them should win the bid in turn so that each of them gains by rotation.

Once the bidding is over, the stock sold is packed in gunny bags by *palledars* (labourers) registered with the APMC. To operate as a registered *palledar*, one has to pay an annual fee of Rs 2 to the APMC. Though there is no contracted obligation, the *palledars* work in informal groups for the Ardhatiyas who were their employers and with whom they had traditional ties as labourers before the APMC was established. As labourers they handle a variety of jobs: unloading, grading, packing, sewing the packed gunny bags, and transporting them on their back to the godown of the *pucca* Ardhatiya who bought the goods.

Payment of sale proceeds to the producer does not follow these operations immediately. There is no on-the-spot cash payment by the buyer (*pucca* Ardhatiya) to the seller (producer), or to his agent (*kuccha* Ardhatiya), or to the *palledar*. Every day, after the sales in the entire *mandi* are over, the *kuccha* Ardhatiyas and their *munims* sit down to record the various entries in their books of account. According to APMC rules, the *kuccha* Ardhatiya must pay the sale proceeds to the producer within 24 hours. Likewise, the rules prescribe that the *pucca* Ardhatiya who bought the produce should make the payment to the *kuccha* Ardhatiya within a week. In reality, however, neither party observes the rules. The *pucca* Ardhatiya takes nearly a month, or often as long as three months, to settle his account with the *kuccha* Ardhatiya, who in turn takes similar or longer time to settle his account with the producer. As neither the producer nor the *kuccha* Ardhatiya considers it advisable to lodge a complaint against the defaulting party and incur its wrath, the APMC officials say that they have not been able to take action against the defaulting parties.

ARDHATIYA-PRODUCER RELATIONS

The Ardhatiya-producer relations described in the foregoing can be classified into three broad categories: market relations (buyer-seller), credit relations (creditor-debtor, through moneylending), and informal relations (both economic and social). These relations are not mutually exclusive. They are in fact so closely interwoven that it may be hazardous to separate them, except in conceptual terms. The discussion which follows should be read keeping this qualification in mind.

The market relations between the agricultural producer as the seller, on the one hand, and the *pucca* Ardhatiya as the buyer, on the other, are essentially the same as that exists between a seller and a buyer in a market where goods and services are openly bought and sold. However, the market for agricultural produce, the *ardhat* trade in particular, being influenced by credit and informal relations, the market relations between the producer and the Ardhatiya are qualitatively different. The three sets of relations running concurrently, one supporting another, have been a continuing feature of the trade, partly because of the seasonality of the produce and partly because of the subsistence nature of the rural economy. Whatever be the reasons, this interlocking of relations has restricted the freedom of the agricultural producer as seller and, in the process, made him dependent on the trader-cum-moneylender.

The market relations have yet another feature. Generally, there is no direct relationship between the *pucca* Ardhatiya and the agricultural producer. Between the two there is at least one more intermediary, the *kuccha* Ardhatiya. Ostensibly, he is the agricultural producer's agent. But, as noted earlier, he often works in league with the *pucca* Ardhatiya, trying to protect the latter's rather than the producer's interests. There are often intermediaries too—the petty village trader and the *faria*, to name a few.

Despite these imperfections and the chain of intermediaries linking the producer with the *pucca* Ardhatiya, the market relations are guided basically by the marketing process and the market forces operating in a given place or situation. Government interventions to regulate the trade are a part of this milieu. While there has been no significant change in Ardhatiya-producer relations as buyer and seller, changes are likely to occur and acquire greater momentum as the character of the market changes either due to government interventions or due to the growth of the agricultural sector from subsistance to sufficiency, if not surplus. Changes are also expected with changes in credit and informal relations, a few examples of which we have already seen and a few more follow.

As stated earlier, credit has always been an integral part of the *ardhat* trade. Almost all Ardhatiyas in the town entered the trade through this route. According to the Ardhatiyas I interviewed at length, the practice in the past was as follows. Once the village trader-cum-moneylender accumulated enough money, he would move to an urban centre in search of new business opportunities. His knowledge of the rural environment, the experience he gained as a trader mainly in foodgrains, and the rural clientele he developed over time facilitated his entry to the *ardhat* trade. He maintained contacts with his cultivator clients and continued to provide them credit on the pledge that they would sell their produce to him.

With rapid increase in agricultural production in the region in the last two or three decades, the business of moneylending has also expanded substantially. This is reflected in the multiplicity of money-lenders and credit agencies. Besides a few licensed moneylenders in the town and the neighbourhood, there are a number of non-licensed moneylenders at the village level. The licensed and non-licensed moneylenders now include as varied a group as village traders, *kuccha* and *pucca* Ardhatiyas, and big and medium farmers. This is in addition to the growing credit from banks and cooperatives.

When I asked some of the agricultural producers as to why they continued taking credit from the Ardhatiyas and private moneylenders instead of from banks and cooperatives, their stock reply was:

> Both are the same for us, except that we are not sure whether the bank or the cooperative will give us the credit. There are various reasons. Firstly, bank or cooperative credit means a lot of *likha-pari* (writing-reading). There are so many forms to be filled in, and we do not know how to do it. Most of us are illiterate. If we have to fill them, we have to take the help of others who know us well and whom we trust. This is a difficult proposition. Bothersome too.
>
> Secondly, banks give credit only for production purposes. If we depend on the bank, who will help us when we need loans to meet social and legal expenses such as those related to marriage and death and fighting court cases?
>
> Thirdly, bank credit is not easy to get. Not all can get it. Those who can, have to spend a lot of time in filling forms and obtaining 'no dues' certificates, visiting various offices, mostly located outside the village, and meeting a number of officials and functionaries. We are told it is not easy to please them. Not only do we have to spend

> time and go through a tortuous process but we also incur some cost. Compared to banks or cooperatives, credit from a moneylender or Ardhatiya is a simple affair. Maybe he gives us less and charges more. But he gives us what we need, at the time we need it, right at our doorstep. We know him, he knows us. And that makes the whole thing simpler.

Notwithstanding such statements, there are occasions when the relation between the Ardhatiya and the producer turns sour. When this happens, it gives rise to disputes. For example, I learned about a dispute between Surendar Singh, a village producer, and Nawal Kishore, a *kuccha* Ardhatiya. According to Singh, he carted a few quintals of wheat to Kishore's *ardhat*. The stock was sold the same day the sales vouchers were made, and copies of the vouchers were given to the concerned parties, including the APMC. As Singh had some urgent work, he left town and returned after a few weeks to claim the money due to him. Singh claims that while the money due to him was Rs 900, Kishore agreed to pay him only Rs 450. Following this, Singh lodged a complaint with the APMC. In accordance with the rules, the APMC secretary called both Singh and Kishore to his office and asked the latter to explain what happened. In Singh's words, Kishore gave the following account:

> Surendar Singh and I know each other for several years. He has been bringing his produce to my *ardhat* for the last five years. After the sale of the stock in question, Surendar left for his home and did not visit my *ardhat* for three months. A few weeks ago I met him in the bazaar and he asked me to give him Rs 450. He said that he needed the money urgently, but forgot to bring the 'entry permit' he collected from the APMC when he brought the stock to my *ardhat*. Since we know each other for several years, I gave him the amount he wanted without any hesitation. However, on my own I recorded the transaction in my books of account, noting that it was a part payment of the amount payable to him. I am prepared to pay him the balance. But now Surendar claims that I did not pay him anything and, therefore, I should pay him the full amount: Rs 900. It is a lie. He is making a false claim. Never before has anyone lodged any complaint against me. My record is clean. Surendar is a liar.

Because of such cases, as also the growing flow of credit from multiple sources, the large and well established Ardhatiyas in the town, particularly the *pucca* Ardhatiyas, no longer prefer to extend credit directly to the farmer. The trend is to route it through the *kuccha* Ardhatiya or the village trader.

The informal relations emphasizing the Ardhatiya's linkage with the agricultural producer are also changing. However, as the farmers participating in the agricultural production system are not a homogeneous group, these relations and the changes in them are not universal as far as the different segments of the farming population are concerned.

For the purpose of this study the agricultural producers are classified in terms of their landholdings, the most critical resource affecting their production and marketable surplus and therefore the nature and degree of their participation in agricultural marketing. Thus viewed, the farmers in the area could be divided into three broad groups: big farmers with holdings exceeding 10 acres, medium farmers with holdings of 5–10 acres, and small and marginal farmers with holdings below 5 acres.

Generally, the big farmers lease out a part of their land on a share cropping basis and cultivate the rest with hired labour. Irrigation in the area being well developed, a major part of the land owned by them is irrigated. Most of them have their own transport (such as bullock carts and tractors). Usually they are not in a hurry to sell the marketable surplus. They store their surplus and keep themselves in touch with the market. They have a feel of the market, and know when they can get a better, if not the best, price. They wait for the market price to rise, and sell the produce when they think the time is ripe. Normally they use their own transport to cart the produce to the town and sell it through one of the *kuccha* Ardhatiyas. Some with contacts with *pucca* Ardhatiyas sell the produce directly to the latter at an agreed but negotiated price. Often the *pucca* Ardhatiya, also a wholesaler, buys the produce but asks the producer to retain the stock with him, in his village, so that he (the *pucca* Ardhatiya) can lift it at his convenience. He does this also to circumvent, if not violate, the RFC rule stipulating that no trader can stock a particular commodity beyond the prescribed limit. Some of the big farmers in the area follow another practice. Like wholesalers, they obtain a licence to stock. They purchase the saleable produce from the medium and small farmers in the neighbourhood and sell it to the *pucca* Ardhatiya at a time of their choosing.

The medium and small farmers are usually in need of money right after the harvest. While the medium farmers have some staying capacity, the small and marginal farmers have none. The medium farmer sells a part of his marketable surplus soon after the harvest and the balance a few weeks or months afterwards, depending on his staying capacity. Generally, for the latter part he manages to get a better price than the one he sells soon after the harvest. He sells his produce through the village trader or the *kuccha* Ardhatiya.

The small and marginal farmers, as elsewhere, have very little to sell and no power to bargain. Because of their compulsions they sell whatever they can even when prices are low. Many of them, being chronically in need of money and always in debt, sell in advance, soon after they have sown the crops, to the village trader or the moneylender-cum-*kuccha* Ardhatiya. The latter or one of his nominees moves from village to village, meets the old and prospective clients, advances credit to them for the purchase of inputs as well as consumption goods, examines the crops they have sown, and assesses from time to time the produce they are likely to get. The fact that he does all this does not mean that he does not exploit them to optimize his profits. This he does by all accounts. But while doing this he also develops informal relations with each client.

Some of the producers, mostly small and marginal farmers, described the nature of this relation as follows:

> Most of us have less than five acres of land. It is the only productive asset we have to feed our families, each consisting of five-eight members. Most of us do not have access to assured irrigation. Our crops and along with them our future depend on the monsoon, which often lets us down. Because we own very little land, we cannot raise or stock seeds from our farms. We have to buy them. As we have hardly any cash to spare, we buy them on credit. We also need credit to buy such necessities as salt, sugar, oil and cloth. We also need money to spend on social rituals and ceremonies like marriage, birth and death. We need loans in kind and cash. Most of these needs are met by our *sahukar*, the moneylender-cum-trader (that is, the Ardhatiya). We know he charges high interest, underweighs the produce we sell him, and also pays less. But he is always there when we need him. We depend on him. He depends on us to run his business.

It is true we have the freedom to sell to government agencies like the RFC or seek the protection of the APMC. We do neither for various reasons. Firstly, we have very little surplus to sell in the *mandi*. If we transport this stock on our own, it will be both costly and time consuming. The Ardhatiya or his agent saves us from this trouble. He takes care of most of our needs. He lends us foodgrains to eat in the lean season. At the time of marriage, birth and death in the family he stands at our doorstep and lends us the money we need without any *likha-pari*. We live from one crisis to another and it is only he who helps us when help is needed most.

We quarrel among ourselves. In the village there is always one or the other *muqudmabazi* (legal dispute) going on. At times the disputes give way to violence because of our fighting among ourselves. There are police cases and we are frightened of them. At such times the *sahukar* again comes to our rescue. He gets in touch with the officials and pleads for us. He talks to the lawyer, presents our case, and persuades him to take it up on our behalf.

Our children are now getting educated. They need employment, regular or casual. We send them to our *sahukar* who introduces them to those who may be in a position to employ them. Till they have found a place to stay, he even lets them sleep in his godown without charging any rent.

There are many more occasions when we need his help and he offers it ungrudgingly. So what if he charges high interest or keeps us in debt so that we may sell our produce to him at a price profitable to him? Frankly, if you ask us, what he charges is really less than the price we may have to pay to obtain loans or assistance from banks or such other agencies. Indeed we don't trust them. We really do not know them. We know only the *sahukar*. He also knows us well. We trust him, he trusts us. No doubt from time to time we have some misunderstanding. But it gets resolved. Both of us belong to the village and we know we have to live together.

This statement illustrates the informal bonds existing between the small agricultural producer and the Ardhatiya. Yet, slowly and subtly these bonds are weakening. The notion of the 'good old days' and time having changed has already started surfacing. This is best illustrated by the following statement from an old Ardhatiya about his impressions of the changes in the Ardhatiya-producer relations over time:

> Those were the days of faith and trust. Both the *kisan* (farmer) and the trader relied on each other. Whenever the need arose we extended support to our clients, the farmers. If they needed commodity or cash for consumption or production we advanced it without much *likha-pari* and they never failed to honour their commitments. The *kisan* was always in need of a loan and we always provided it. Besides regular payment of interest, he brought and sold his produce through us. The times have now changed. We *beoparis* have become greedy. We are not satisfied with having limited but trustworthy clients. We are always looking for more and more profits. The *kisan* has also become dishonest. He takes money on credit from us but sells the produce to someone else. We now mistrust each other.

This response is indicative of the decline in the patron-client relationship. Slowly but surely it is giving way to contractual relations. Many Ardhatiyas reported this. They felt that this was happening because of growing state intervention and the politics of election. This may be true. But it is also happening, in my view, because of the development process under way.

CONCLUDING OBSERVATIONS

The changes noted in the earlier discussion have yet to gather momentum. They are too slow to be visible, except perhaps to the discerning eye. Nor has there been any perceptible change in the basic character of the trade. Most of the features described in the section on the history of the pattern and structure of the *ardhat* trade exist even now. The fact that the basic character of the trade has remained unaltered suggests that continuity rather than change has been the dominant feature of the trade, if not the dominant ethos of the actors participating in it.

Changes have occurred only during the last two or three decades. These changes, however slow or imperceptible, have been due largely to the following developments in the country encompassing not *ardhat* trade alone but the society at large:

1. The general social and economic development, with emphasis on planning and planned interventions to accelerate the rate of economic growth, including the development of the agricultural

sector, modernization of society in general, and democratization of the polity.
2. The social, political and economic processes (both macro and micro) these efforts have set in motion.
3. The growing need for income-earning opportunities and the varying responses of different social groups depending on their capability to harness them.
4. Government interventions to regulate the grain trade and to participate in it through legislation, procurement, financing, and changes in the trade or market structure.

While all these have influenced the three sets of Ardhatiya-producer relations described earlier, including in some ways the structure of *ardhat* trade, I shall restrict my observations to the last two. Take, first, the structure of the grain trade. Over time and following the various government interventions, the structure as it exists today looks highly complex. The actors in the structure include numerous and diverse groups, such as agricultural producers, itinerant traders, peddlers, village cooperatives, 'floating' Ardhatiyas or *farias*, *kuccha* and *pucca* Ardhatiyas, wholesalers, retailers, the APMC, the RFC, fair price shops, mills, and presses.

There are other complexities. The *pucca* Ardhatiyas are not only commission agents but also wholesalers. They function as commission agents of their own wholesale firms and those in other towns. They are agents and stockists of the RFC. They have to register their firms with the RFC, obtain licences from it, and work under its supervision. Without these licences they are not authorized to buy and sell on behalf of non-local parties, nor can they despatch agricultural produce outside the town or stock on behalf of outside parties.

One consequence of this multiple control, as reported by the town traders themselves, is collusion between the Ardhatiyas and the RFC employees to beat the system and share the booty. There is no documentary evidence to validate the statement, but the perception that such is the case is widely prevalent among the traders. The perception is strengthened by the following case reported by one of the traders:

> The RFC has a store-cum-office in the town. It is owned by a *pucca* Ardhatiya who happens to be an agent of the RFC itself. He receives rent as well as commission from the RFC. Indeed, he and four of

his friends constructed the building on the request of the RFC and on the assurance that the RFC itself would hire the building. Can you therefore blame us if in the local gossip the implication of such close links often figures?

Collusion to evade measures which are too demanding and yet not easily enforceable is not a new practice. Nor is it confined to Ardhatiya-RFC collusion. What is being emphasized is not the existence of collusion *per se*, but the processes which structural changes often set in motion.

The search for income-earning opportunities in an area where opportunities are few, or limited to certain parts of the economy, can manifest itself in various ways. The Banias in the town seem to view *ardhat* trade as the focal point in their search for income-earning opportunities. There could be several reasons for this. First, their traditional social values emphasize that their career lies in trade. Second, they would have gained experience and contacts over generations as traders and moneylenders. Third, they see the spurt in agricultural production in the region and the national concern to promote the growth of the agricultural sector, and therefore view it as the most promising field for beginners like them. Fourth, *kuccha ardhat*, the entry point to the trade, is a low-risk, low-investment venture and therefore not difficult to enter.

Even though the mortality rate in *kuccha ardhat* in the town is high, it keeps attracting new entrants—almost invariably Banias—every year. According to the APMC records, every year 20–25 *kuccha ardhat* firms are registered in the town, of which 10–15 get out of business either in the same year or one or two years later. Usually those who manage to survive the first year keep changing the name of the firm—dissolving an existing firm and registering in its place another with the owners or partners remaining unchanged.

Usually a new entrant to the trade adopts either of the following two ways to get acquainted with the trade and become a successful Ardhatiya.

1. If he has very little or no money to invest, he tries to persuade the owner or the partners of a well established *ardhat* firm to employ him as one of its functionaries with or without any remuneration. If he succeeds, he tries, first, to learn the tricks of the trade and simultaneously develop contacts with the producer-clients. Thereafter, he starts sending feelers to the producers that

he is one of the partners of the *ardhat* firm with which he is associated. Occasionally he also bypasses the firm by buying and selling the produce of some clients as an independent Ardhatiya. Gradually he develops links with the producers. When he feels sufficiently confident that he could start his own *ardhat* business, he leaves the firm and floats his own firm to join the body of registered Ardhatiyas.

2. If he has sufficient capital to invest but very little knowledge of the trade, he requests one of his relatives or caste members who is well established in the *ardhat* trade to let him or his brother or son work as an apprentice. Once he or his nominee gains sufficient knowledge of the trade, he establishes his own *ardhat* firm in alliance with the owner or the partners of the firm where he or his nominee worked as an apprentice.

Most of the new entrants opt for the first alternative because they have very little capital to invest. The amount invested varies from individual to individual but it is seldom more than Rs 10,000. They operate with as low a commission as 1.5 per cent and manage to get incomes ranging from Rs 100 to Rs 200 a month. As this is too small a sum to survive on and to participate in the trade on a full-time basis, they generally try to expand their capital base and business by choosing one or more of the following alternatives.

1. Transfer to the producer as much of his risk and costs as possible.
2. Keep the *pucca* Ardhatiya in good humour and persuade him to settle the accounts as quickly as possible. The shorter this period the better, because he (the new entrant) can rotate the capital faster and thus handle more business.
3. Do all this but keep transferring a part of the capital to other ventures.
4. Failing all these, accept defeat and quit.

Almost invariably the *kuccha* Ardhatiya tries to shift to his producer-client the problem he might face due to delayed payment by *pucca* Ardhatiyas. By withholding payment to the producer, or by paying it in instalments, he also seeks to strengthen his hold over the producer. He makes the producer more dependent on him by compelling or inducing him to meet him several times. This helps him to have frequent interaction with the producer and develop informal ties with him.

Deferred payment is also a form of credit which the Ardhatiya manages to get free of interest. To the extent he gets such credit, he is able to expand his capital base without any additional cost. He may invest this capital to (*a*) expand his *ardhat* business, (*b*) transfer a part of it to other economic pursuits such as moneylending, or (*c*) develop and diversify his productive resources. Generally the Ardhatiyas, both *kuccha* and *pucca*, try to follow one or more of these options. Some succeed. Some fail and get out of the business.

It is arguable whether the Ardhatiyas, usually Bania by caste, could be viewed as entrepreneurs. Scholars who have debated the issue hold different views. Some consider them entrepreneurs, some as agents of development, some as agents of underdevelopment, some as agents of stagnation. While I do not wish to join the issue here, the following points about the Ardhatiyas of Sikandra Rao may have some bearing.

1. The Ardhatiyas are a distinct group occupationally as well as socially—the latter in the sense that virtually all Ardhatiyas in the town belong to one or other subcaste of Banias. This was the case in the past, remains so even today, and is likely to be so for years to come.
2. In small towns like Sikandra Rao, the Bania dominates the commercial and trading activities, both traditional and new. The former includes products of mass consumption like foodgrains and cloth. The latter includes relatively modern products originating from recent development efforts and new ways of life, like electrical goods, toiletries (like soap, talcum powder and cream), pump sets, agricultural equipment and accessories, fertilizers and other agricultural inputs. It is true that the new goods and services do not represent a high level of technology, that the trading operations are not of a high order, and that the risks involved are low. Yet the fact remains that, in the town, only the Banias have taken advantage of the growing development opportunities. Some of them, particularly the well established Ardhatiyas, have also entered the field of industrial production, though on a small scale and involving relatively low technology, such as setting up rice and flour mills, brick kilns, and rubber reprocessing units.

Most Ardhatiyas in the town have followed nearly the same pattern of growth. Usually they began their careers as petty village traders, combined this activity with moneylending, moved to the town after they accumulated enough savings, joined the grain trade as *kuccha* Ardhatiyas, and, depending on luck, rose to the level of *pucca* Ardhatiyas. As we have seen, only five of the present 25 Ardhatiyas in the town have become *pucca* Ardhatiyas. Throughout their careers they combined moneylending with trading. Until recently they did this openly. The *kuccha* Ardhatiyas do it even now. The *pucca* Ardhatiyas now prefer to do it through other intermediaries, for reasons stated earlier.

As and when the Ardhatiyas settled down in the *ardhat* trade, they started diversifying their investments and interests. Before the town became an important grain marketing centre in the area, they invested in orchards and real estate, and retailing in cloth and groceries. The present trend is to get into agro-processing industries on a small scale and become distributors of agricultural inputs and agents of private but large manufacturing industries.

Social continuity amidst economic change, which is slow but steady, is still the dominant theme of the social and economic life of the Ardhatiyas (Bania traders) in the town. Whether the rate of change will get accelerated enough to disrupt this theme, time alone can tell. But this study does suggest the need for more intensive efforts by social scientists to study the place and problems of small towns from various perspectives, including the one I have presented in this article.

REFERENCES

Bayly, C. 1975. *Rulers, Townsmen and Bazars in North India.* London: Oxford University Press.

Buchanan, D. H. 1934. *The Development of Capitalist Enterprise in India.* New York: Macmillan.

Census of India. 1971. *District Census Handbook: Aligarh.* Lucknow: Director of Census Operations. Uttar Pradesh.

Frykenberg, R. E. (Ed.). 1969. *Land Control and Social Structure in Indian History.* Madison: University of Wisconsin.

Gadgil, D. R. 1959. *Origin of the Modern Indian Business Class.* New York: Institute of Pacific Studies.

Gupta, Khadija. 1975. *Politics in a Small Town.* Delhi: Impex India.

Jain, L. C. 1927. *Indigenous Banking in India.* London: Macmillan.

Mitra, A. 1974. *A Functional Classification of Indian Towns.* Delhi: Institute of Economic Growth.

Nevill, H. R. 1926. *District Gazetteers of United Provinces of Agra and Oudh,* Vol. VI. Lucknow: Government Press.

Ray, R. K. 1982. 'The Bazar: Changing Structural Characteristics of the Indigenous Sector of the Indian Economy, 1914–1947', paper presented at the Seminar on Dynamics of Indian Business. Ahmedabad: Indian Institute of Management, mimeo.

Singer, Milton and Cohn, Bernard (Ed.). 1968. *Structure and Change in Indian Society.* Chicago: Aldine.

Srinivas, M. N. (Ed.). 1977a. *Dimensions of Social Change in India.* Delhi: Allied.

——. 1977b. *Science, Technology and Rural Development in India.* Poona: Gokhale Institute of Politics and Economics.

Thornton, E. 1954. *A Gazetteer of Territories under the Government of the East India Company and Other Native States on the Continent of India.* London: W. H. Allen.

Tripathi, D. (Ed.). 1984. *Business Communities of India: A Historical Perspective.* Delhi: Manohar.

10

Towards a General Theory of Urbanization and Social Change

HARSHAD R. TRIVEDI

INTRODUCTION

This paper[1] is a continuation of my earlier paper entitled 'A Theory of Urbanization in India' (1971). The purpose of that paper was to present a broad theoretical framework and a typology of what are called 'semi-urban pockets' (hereafter SUPs), based on data derived from India. Two broad general hypotheses were presented there, but

[1] This is a revised version of a paper submitted to the Eleventh International Congress of Anthropological and Ethnological Sciences held at Chicago in August-September 1973. I thank K. Venkataraman of the Indian Institute of Public Administration, New Delhi, for giving useful suggestions on the first (1968) draft of the paper. The concept of SUP on the basis of which this paper is developed was evolved during my tenure as Deputy Director of Cultural Anthropology at the National Institute of Community Development, Mussoorie in 1963.

their context was specific as they referred only to India. There was, however, scope for putting them to test in a few other countries. I have tried to do this here in a limited manner on the basis of secondary information, and have also evolved a new complementary hypothesis. This third hypothesis, presented later, adds a universal dimension to the previous two hypotheses. The main thrust is towards evolving a general theory of urbanization in the light of the analysis by Sjoberg (1967a, 1967b). A good deal of material used by him in clarifying certain theoretical controversies concerning urbanization as a process of change and urbanism as a way of life is used in expounding the theory set forth here. In this process of theorizing, I have established an interrelation between the three hypotheses.

A theory of urbanization is basically a theory of social change, implying that urbanization is both the cause and effect of social change. As indicated in my earlier paper (1971), a few decades ago the tempo of change in India was not high, and therefore society retained a considerable degree of differentiation between its urban and rural macro-social structures. The introduction of technological advances in the systems of communication, industry, education, and so on, in the last four decades has created a state of partial modernization in India. As a consequence, the two old polar structural realities of 'rural' and 'urban' have given birth to a third reality—named SUP—with varying shades of urbanism. The present rapid, widespread and interpenetrating communication patterns and the role of agents of change make it extremely difficult to accept with reason the dichotomy of pure rural and urban settlement types. The focus of this article in reality and essence is on the changes taking place in human behaviour in urban areas subject to the influences of multiple forces—geophysical, historical, economic and socio-cultural.

The compulsion to forge secondary relationships in place of primary ones with different types of people for legal, political, economic and intellectual pursuits, indicates a centrifugal urge in the choice of migration and settlement. To begin with, villagers went to towns to earn a livelihood. However, in the new environment these migrants tend to live longer in slums and semi-urban pockets. Similarly, the people who settled for generations in old settlements in large towns or cities are desirous of protecting their primary and quasi-primary social relationships in the midst of a complex network of secondary relationships all around. They tend to remain confined to semi-urban pockets, indicating a centripetal urge. It is appropriate to use the term SUP for such a composite new social reality.

As the concept of SUP is crucial to my analysis, it is necessary to explain it at the outset. My book, *Urbanism: A New Outlook* (1976) gives an idea of the development of the concept, which I shall summarize here.

It seems reasonable to assume that SUPs range in population size between 5,000 and 20,000. Recent observations show that the optimum population size may increase during meso-level growth (which may be called semi-urban proliferation) as well as during macro-level growth (which may be called semi-urban periphery). There are 'rural SUPs' and 'urban SUPs', both having less of rural and more of urban characteristics, and interpenetrating into hinterland villages on the one hand and urban towns on the other. The rural SUPs in India cover market towns, temple towns, hill towns, sea ports, labour colonies, and villages within towns and cities. The urban SUPs cover industrial townships, cantonments, refugee colonies, old and new suburbs, old settlements of towns, and railway colonies. A combination of economic, geographical, political, historical, social and administrative factors influence the conception of this typology. Less emphasis is placed on such criteria as size and density of the population and land use, which are considered important in the national census, physical planning and socio-economic surveys. Research is required to establish precise criteria for identifying the genus SUP, the species rural SUP and urban SUP, and the various sub-species. For practical developmental purposes, all SUPs can be further identified either as 'generative' or 'parasitic', depending on their potential for viability and growth or stagnation. The concept of SUP, however, highlights the fact that under the impact of technological development and industrialization, traditional society in India is being reorganized at the micro, meso and macro levels into a heterogeneity of types and models.

The typology of rural and urban SUPs is presented in illustrative form in Table 10.1. As mentioned earlier, these are small and medium towns with a population of between 5,000 and 20,000 persons.

There is no general agreement among social scientists on a definition of 'urban', let alone 'semi-urban'. However, the term 'semi-urban' refers here either to the lack of some important urban characteristics or to the partial growth of a number of such characteristics singly or in various combinations. The important common characteristics of all SUPs, irrespective of sub-types, are as follows (and the term 'semi-urban'

Table 10.1 ***Illustrations of SUP Sub-Types***

Rural SUP	*Urban SUP*
1. Market town, bazaar town	1. Old cantonment, old police lines
2. Feudal estate town, temple town	2. Old and new suburb in a large town or city
3. Hill station, tourist town	3. Old settlement in the centre of a large town or city
4. Coastal town, riverside town	4. Railway housing colony for lower class railway employees
5. Railroad junction town, transportation nodal town	5. Colony of refugees and/or immigrants
6. Urban village, labourers' colony	6. Small industrial town, university town

refers to those communities and their parts which possess these characteristics): (*a*) underdevelopment of economic and technological spheres, (*b*) lower level of secondary social relationships as compared with primary and quasi-primary relationships, (*c*) differentiation and discontinuities in small group or individual behaviour, social relationships, institutional linkages, integrative forces and social systems, (*d*) a low level of civic administration and of citizens' political participation in local bodies, (*e*) the lack of new supporting institutions to harmonize discontinuous and weak linkages for integrated community development, (*f*) the slow and sluggish emergence of rudimentary forms of class structure, and (*g*) a low level of occupational and social mobility.

The term 'pocket' suggests an area, population, community and culture-complex with a role in the larger social whole of which it is a part. Which attributes of the whole are relevant to the pocket, and which attributes of the pocket are relevant to the whole is difficult but not impossible to spell out. A scale of attributes and their significant combinations will have to be framed not only for locating SUPs in the form of a basic typology, but also for reclassifying residential and industrial areas as they shift from one type of SUP to another over time. As mentioned earlier, the term 'pocket' is inadequate to cover the entire range of new behavioural attributes of social change, because sometimes semi-urban situations are found on the periphery of a large town or city in the form of loose population proliferation and changing social pockets. Thus one can talk of 'semi-urban periphery', 'semi-urban proliferation' and 'semi-urban pocket', irrespective of the socio-cultural nature, size and distribution of the population.

DEFINITION OF BASIC TERMS AND CONCEPTS

To begin with, I shall clarify my general stance and define some basic terms and concepts. My preoccupation with the concept of SUP impels me to exclude from present consideration certain issues, which others, for their own reasons, may not find significant for the study of urbanization. I cannot avoid becoming selective in this matter. Such limitations in theorizing occur in most attempts of this kind and will, I hope, be considered acceptable and valid.

In its composite sense, the term urbanization is equivalent to 'urban' as discussed by Popenoe (1969: 70). However, the terms city and urban should not be considered synonymous, because by convention and in reality they appear to be both inclusive and exclusive of one another. In my definition of SUP, the term urban carries (*a*) similar connotation (namely, structure and function of urbanizing communities), and (*b*) similar denotation (namely, causes, conditions and consequences of urbanizing processes). My contention, therefore, is that the three identifiable social realities—urban, rural and semi-urban—co-exist in India and other underdeveloped societies today. They call for an appropriate use of the term community, which assumes different characteristics under different conditions. According to Popenoie, the focus of the community is on 'inter-organizational and inter-institutional relationships and integrative mechanisms at the level of "localized" social interaction within a society. This is sometimes referred to as the 'residual' notion of community, or what is left when primary functions and activities of functional and sub-locality groups are analytically factored out' (ibid.). I consider the first part of this definition applicable to urban and rural communities. The second part (that is, the residual notion) indicates the existence of a third reality, which is a part of the entire community structure in a society, but which is primarily concerned with functions and activities of the people living in SUPs and localized sub-communities.

This does not mean that all SUPs are discarded or left-over realities in a larger national community. These sub-communities may be qualitatively different from one another. The differentiation takes place on account of varying combinations of diverse characteristics of group relationships (such as endogenous, exogenous, latent, manifest and symbolic or real) in terms of social life in action. What needs to be emphasized, however, is that most SUPs may be devoid of operational mechanisms of identity, such as viable political process, clear class stratification, equitable distribution of power and expressive social integration, if they are also not devoid of a public market-place, system of communication and transportation, and real and symbolic wholeness. This can be ascertained from the tentative typology proposed in my 1971 paper and reproduced earlier.

The city cannot be considered the ideal type of urban social reality on account of the coexistence of culturally homogeneous but otherwise heterogeneous social and natural areas and the existence of semi-urban reality as an outcome of the interpenetration of rural and urban elements. In fact, the relationships of natural and social areas (Park 1952) with non-materialistic content are complementary to the materialistic nature of the city at the present time in human history. This is evident in the study of divergent cultures, such as those of the US, Africa and Asia (including India). If this line of thinking is of any value, it could be tested in societies with capitalist and socialist polity as well as traditional and modern ethos.

The terms 'traditional' and 'modern' could be applied to many contemporary societies, but the thesis that various modern communication networks positively effect unprecedented changes in different societies, such that they give rise to SUPs, calls for understanding these two terms with reference to their coverage of social reality wherein the emergence of the third SUP reality is feasible. Rudolph and Rudolph (1967) have done an admirable job in providing the composite concept of 'modernity of tradition', which in simple terms suggests that the phenomena of modernity and tradition are relatively different but interpenetrating realities affecting the organization and structure of societies at a point of time. My contention is that it is mainly in the process of political socialization that this is happening in India today. It is likely that the reverse of this process (namely, traditionalization of modernity) is occurring, at least in the sociological sense, in developed countries. But this is beside our main concern. In fact, whatever be the degree of dichotomy found in most societies with regard to modernity and tradition, it is not qualitatively the same everywhere. For instance, according to Rudolph and Rudolph, 'the literature focusing exclusively on so modern a society as America tends to contradict the notion that tradition and modernity are dichotomous' (1967: 74). However, the fact remains that it is difficult to arrive at precise definitions of modernity and tradition in general.

In the context of urbanization, however, definitions of tradition and modernity can be provided by accepting the dictionary meaning of these terms. According to Fairchild (1955), the term 'modernity' is derived from 'modern', which refers to those ideas, practices and things with materialist overtones that are not ancient or remote and those that pertain to present and recent times. It refers to the quality of being modern, in a more or less material and symbolic sense. On the other hand, the term 'traditional' refers to adherence to tradition as authority, especially in the matter of religion which includes non-materialistic elements of philosophy, ethics, morality, and so on, derived from the past in the form of customs, beliefs, rituals and practices. It is evident that with the contemporary revolution in audio-visual information systems and spatial communications in today's world, there is hardly any society that can be called purely traditional or purely modern. In this context of the study of urbanization, it is important to retain the level of abstraction contained in the dictionary meaning of these two terms, so that a cultural bias is not introduced in the development of a general theory of urbanization and social change.

As suggested by Sjoberg (1967b: 159), the approaches to the study of urban phenomenon fall into two broad categories—materialistic and non-materialistic. The non-materialistic approach covers the role of social and cultural values, attitudes and behaviour of people as determinants of urbanization. The materialistic approach gives priority to the external environment, land-use pattern, population structure, and so on. The non-material aspect of urban life is derived from the natural will or *gemeinschaft* elements while the material aspect derives from the rational will or *gesellschaft* elements to be found in human values, attitudes and behaviour. The non-materialistic approach focuses attention on 'urbanism' largely as a way of life, both in terms of values, attitudes and behaviour. As defined by Meadows and Mizruchi:

> It is a pattern of existence which deals with (1) accommodation of heterogeneous groups to one another, (2) a relatively high degree of specialization in labour, (3) involvement in non-agricultural occupational pursuits, (4) market economy, (5) an interplay between innovation and change as against the maintenance of societal traditions, (6) development of advanced learning and the arts, and (7) tendencies towards city-based, centralized governmental structures. Urbanism refers to a number of values which can be intuitively

> perceived as a whole but which have yet not been adequately dissected by social scientists (1969: 2).

Likewise, the materialistic approach to urbanization refers primarily to the processes, causes and effects of materialistic changes leading to the concentration of people in an area as well as the results of the same affecting physical conditions in expanding human settlements. Also, as Meadows (1969: 14) states: 'Urbanism depends upon the appearance and growth of an economic surplus.' According to Doxiadis (1967), the essence of the city consists of five basic elements: nature, man, society, shells and networks covering both the material and non-material aspects of social existence.

Qualitatively, 'urbanization refers to the processes by which (1) urban values are diffused, (2) movement occurs from rural areas to cities, and (3) behaviour patterns are transformed to conform to those which are characteristic of groups in the cities' (Meadows and Mizruchi 1969: 2). From this point of view at least, the non-materialistic and materialistic interpretations are complementary to one another. It is, therefore, considered proper to look upon them as one whole phenomenon in the broader frame of urbanization which conventionally includes both urbanism as a way of life and urbanization as a process. Meadows (1969: 16–18) comments on some scholars who emphasize special aspects of urbanization in terms of theories such as (*a*) time honoured, standardized, mechanized, packaged, priced and merchantized way of life, (*b*) dichotomic moralism of primitive, rural and urban cultures, (*c*) a heroic theory of the city which looks upon it as an abstract culture-hero creating, sustaining and elaborating human values, and functioning as a source of power influencing the hinterland and the space beyond, and finally, (*d*) an entrepreneurial theory which looks upon the city as an important agent of collective enterprise—dynastic, ecclesiastical, military, political, industrial, and so on. Obviously, this does not give a balanced view of urbanism or the city as such, and we should naturally look for a higher level of abstraction.

ABSTRACTION NEEDED FOR THEORIZING ON A GENERAL PLANE

Broadly, Sjoberg considers that the main purpose of urban sociology is 'to understand and predict, by whatever theoretical or methodological tools are available, the social and ecological structure of cities or actions of their inhabitants' (1967a: 159). This statement has two main implications. First, it admits the possibility of more than one set of theoretical and methodological tools for the study of urban society, depending on the purpose of the study. Second, it equates urban society with the city in a covert sense, such that urban society is not possible outside the framework of the city. This position seems alright so far as it admits more ways than one to study urbanization within the ambit of the city. However, the level of abstraction for theoretical purposes can be raised while studying a specific urban society, with the expectation that it can be applicable with varying degrees of modification to most of the contemporary societies passing through diverse processes and stages of urbanization. To be able to do this, we must set aside the concept of the city, particularly as defined by the materialistic approach which considers urbanization as a dependent variable external to the actors and their way of life in a town. For, apart from the material conditions that make a city, there are non-material conditions which also make an urban community, that is, as an independent variable internal to the actors whose values and motivations are central to the results of their actions. These are two sides of the same coin. However, the non-material side of the coin is amenable to a higher level of theoretical abstraction, such that it may encompass urban and urbanizing societies in various types of traditional or modern contemporary cultures. In other words, it is possible that the two differentiated phenomena—urbanism and urbanization—can be separated on a conceptual plane and one of them be treated on an abstract level so that the universal process of interpenetration of rural and urban elements and social characteristics, irrespective of city as the so-called ideal urban community, can be understood well.

The concept of SUP has been evolved by setting aside the classic concept of 'city' from the point of view of variety and totality of community organizations in contemporary global society. For the construction of the basic ideal types of cities given by Max Weber cannot be complete without correcting the inherent dichotomy of the concept, as pointed out by Murvar (1969: 53). The oriental city of today has become juridically, constitutionally and materially distinct from the village. It is no more organic or homogeneous in the sense in which occidental cities were in the past, nor is it devoid of an egalitarian political ethos. For example, with the abolition of 'privy purses' and other privileges of the Indian princes, the feudal monopoly of economic, religious, military and political power in towns and cities within erstwhile princely territories does not exist today. Similarly, in the technically advanced American society, the fast changing patterns indicating how populations are dispersed in standard metropolitan areas with an increasing rate of suburbanization make the concept of the homogeneous city obsolete (see Tarver 1958: 195–205).[2]

RELEVANCE OF EARLIER PROPOSITIONS AND HYPOTHESES

It is my contention that, except in parts, the city today cannot be understood adequately from the viewpoint of social structure and organization.[3] However, there is no denying the fact that the census enumerations and such other surveys throwing light on the broad characteristics of a city are important. As Sjoberg puts it,

> the expansion of technology, notably industrialization, not only gives impetus to urbanization but is itself spurred by the growth of cities. Also, definite ties exist between technological advance and the dominant ideology and power structure. A society's value orientation, or ideology, determines to a marked degree the manner in which social power is applied (1967a: 178).

[2] Here we may note what Gibbs and Martin (1969: 158) have to say about the prerequisites of urbanization: 'The value system of some societies may in fact favour a high degree of urbanization, but there is no particular set of values that is a sufficient condition for a high degree of urbanization. It makes no significant difference whether the population professes socialism or capitalism, liberalism or conservatism, Buddhism or Free Methodism, for, if a high degree of urbanization is to be maintained, widely dispersed materials must be requisitioned, and this can be accomplished only through a complex division of labour and technological efficiency.'

[3] A similar opinion is expressed by Orleans (1969: 104) while discussing the concepts of 'natural areas' and 'social areas': 'There are undoubtedly spatially distinct and socially differentiated population aggregates in the urban milieu which fails to generate either associations based upon spatial contiguity or associations predicated upon commonalities of interest. Such segments of the urban conditions represent significant social (or perhaps, more correctly, asocial) form. For such populations, Park's concern with the problem of order is critical. But it would be a serious error to focus attention primarily upon such populations, considering them representative of urban conditions as a whole, and thereby obscuring a conception of urban social organization.' A certain ambivalance is reflected in the latter part of this statement, with the positive contention that natural or social areas in a city are somewhat distinct realities, like SUPs, when treated separately for socio-cultural studies of cities.

A similar stance was taken by Wirth (see Reiss 1964: 63) when he said that we are not likely to arrive at any adequate conception of urbanism so long as we identify urbanism with the physical entity of the city.

Unlike the twelve propositions of Wirth on this subject (see Morris 1968: 16) which look as if they emerge from observations made within the purview of urban phenomenon, this theory, propounded in the shape of three interrelated hypotheses, appears to emerge from the outside. In the process, it helps identify the character of urbanism and urbanization, leading to three different but not distinctly separate realities to be found in human society in the present technological and atomic age. I must admit that this theory does not answer any specific questions of practical import, except to indicate in a general way the outcome of the tempo of change and the direction in which human societies are heading under the impact of technological advancement and an ever-growing network of communications.

The neo-evolutionary hypothesis of urbanization based on the impact of technology, as postulated by Sjoberg (1967b: 179), is no doubt refreshing. It says, 'as industrial urbanization proceeds and as the technological environment becomes increasingly complex, the possibility of evaluating one's external environment in a variety of ways is concomitantly enhanced.' This converse hypothesis is at the root of my thinking. It says that the internal social and cultural environment, as conceived in the spirit of the non-materialistic approach, is the key to identifying the third, the semi-urban, reality under varying degrees of socio-cultural development, shifts and changes. Just as the preliterate and modern societies display a wide variety of values in the familial, political and religious spheres, so also one can discern a great contrast of values, ideas and beliefs in industrial-urban systems. All individuals in a society (traditional or modern) do not value positively and with equal emphasis complex tools and scientific knowledge. Reissman, among others, has illustrated this vividly (1958: 379–90). Surprisingly, this appears to be the case with the hippies' aversion to materialism in industrially advanced societies, as well as one finds have-nots disenchanted with parochial, religious and ethnic groups in the developing societies. These non-conformist groups and individuals symbolize elements of SUPs in either latent or manifest forms, depending on the growth of industrialization and urbanization in a country.

PRESENTATION AND ANALYSIS OF THREE INTERRELATED HYPOTHESES

It should be clear from the preceding discussion that, in terms of manifest and latent functions, the concept of 'city' with the dominant materialistic (manifest) aspect is separable from the concept of 'urban' with the dominant non-materialistic (latent) aspect. The concept of 'rural' with its origin in the traditional method of community organization based on the tribal-folk culture continuum provides an organic link between the materialistic and the non-materialistic way of life. The concept of SUP provides a link between the urban and the rural way of life.

It appears that SUPs in India are like some American suburbia exhibiting demographic, ecological and organizational characteristics of the middle class, identified at a given point of time by the prevailing standards of that society. In other words, SUPs in India, like suburbia in the US, are found not only outside the fringe of a metropolitan city but also inside the core of its limits (Dobriner 1958: xv–xxi). Besides, they are found even in the hinterland of a metropolitan region. In brief, as an end product, it is the style of interpenetration of manifest (materialist) and latent (non-materialist) social elements and functions in a particular country that determines the origin and nature of SUPs or suburbia, a third social reality with distinct characteristics indicating a combination of rural and urban elements and styles of life.

The theory presented here tries to provide an answer, with the help of hypothetical propositions, to the three basic questions of 'what', 'how' and 'why' of urbanization in general. At the outset it may be noted that these propositions are multivariate in nature, as the inductive method of reasoning has been used for the purpose of theorizing. The first hypothesis suggests 'what' it is that we as sociologists and anthropologists need to keep in mind while developing a rational outlook on the study of urbanization. The second hypothesis says 'how' the urbanization process comes about, and indicates the general macro-processes of change that lead to the emergence of a third, semi-urban or suburban reality in the fast-changing world of today. The third hypothesis explains 'why' the new reality takes birth in contemporary societies, irrespective of the stage of their social and economic development.

Hypothesis 1

The external conditions and internalized values of rural people determine the social and spatial aspects of urbanizing sub-group settlements which are amenable to study not only from the non-materialistic and materialistic points of view but also with regard to fusion of the two into a semi-urban reality.

It would be evident from this hypothesis that the concept of SUP is a viable means of attaining a higher level of abstraction in studying urbanization. The empiricists who have established different schools of thought and theories by generalizing from painstakingly gathered data on specific aspects of the city may not venture to undertake such an exercise. However, they would appreciate that this attempt at theorizing highlights the utility of a conceptual framework on the basis of a tentative bivariate cross-settlement typology for identifying the salient characteristics of the contemporary urban phenomenon.

The most important features of urbanization for our consideration are the external conditions and the internalized values of rural and urban born people. The size-density model derived from Durkheim represents the external conditions giving a specific shape and direction to the social morphology of a city. As Durkheim states:

> The social substratum differs according to whether the population is large or small, and more or less dense, whether it is concentrated in cities or dispersed over the countryside, how cities and houses are constructed, whether the area occupied and the society is more or less extensive, and according to the kind of boundaries that delimit it (see Hauser and Schnore 1966: 11).

The latter part of the statement (namely, how cities and houses are constructed) is of crucial importance. The concept of SUP is inherent in it, though not stated explicitly. The density and size of the area occupied by ethnic groups in a society no doubt determine the latent differences in the social substratum. As may be seen from my earlier papers (1969 and 1971), the size of the population in an SUP type of settlement may vary from 5,000 to 20,000, concentrated in a limited area, and it may be part of a large city or may just exist by itself (in a country like India) as a small or medium size town in the hinterland. It may not be dispersed in the form of a group of villages or in any other haphazard way. The nature and form of any SUP represents a combination of materialistic and non-materialistic aspects of human life, conditioned ecologically to a minimum and a maximum size limit, and to a form of community organization marked by a mixture of primary, quasi-primary and secondary sets of relationships. In brief, the concept of SUP represents the fusion of qualitative and quantitative aspects of urbanism and urbanization. This hypothesis is the first step in the evolution of a fuller theory of urbanization.

It will be clear from a subsequent appraisal of different schools of thought that most of the previous scholars had given high priority to the materialistic aspects of the city for evolving hypothetical propositions and theoretical generalizations, and kept the non-materialistic aspects in the background. The foundation of these propositions appears to be sound and based on concrete facts, but for that very reason the whole approach becomes rigid and consequently of partial significance in unravelling the total nature of the urban phenomenon. For a general theory to be relevant, it should rise above the particular time-and-space-bound propositions, be applicable to universal situations, and become relatively value-free.

Hypothesis 2

The present-day urbanization is the outcome of the impact of continuous and instantaneous forces of change to which human societies have been subjected, and on account of which they have assumed a variety of new forms and structures.

Briefly, this hypothesis tells how and why the emergence of the third social reality in the form of SUP settlement types is inevitable. The main emphasis here is on historical determinism. However, the instantaneous forces (such as war, political revolution and natural catastrophe) may also lead to unprecedented changes in the structure and organization of an urbanizing society and its settlements.

Our focus lies now in knowing how SUP settlements come to exist in latent or manifest forms in the fast changing world of today. As the historical forces of change are constantly impinging on various aspects of life, a general reaction of many people is regressive dependence on traditional values, attitudes and old patterns of behaviour, in contrast to aggressive behaviour of some sensitive people in favour of new changes. The historical forces lead to internalization of beliefs and practices. Generally, these forces represent continuities in primitive technology, a preponderance of primary relationships, subsistence economy, including household industry and agriculture, untamed geographical and ecological resources, undiversified social organizations, static religious and administrative institutions, a preponderance of ascriptive values over universalistic ones, the hold of traditional political elites on politico-legal processes, pride in the old material culture and artifacts along with the predominance of caste, ethnic groups and incipient class structures. Similarly, the impact of the diffusion of cultural traits from other societies in the fields of philosophy, religion and attitudes towards sex, housing, dress, recreation, and so on, may create chain reactions in other spheres of social life in rural as well as urban areas. All these historical forces impinge heavily on human life and settlement patterns, though they may and can be altered by the sudden and powerful thrust of instantaneous forces.

The instantaneous forces which may lead to unprecedented changes in the structure and organization of a society and its settlements, as indicated, are war, political revolution, natural calamity and scientific explosion. These changes are translated in the adoption of and adaptation to innovations, new machines, tools and techniques of doing things, and thereby achieving new goals in space travel, nuclear power, and so on. Other forces leading to rapid and instantaneous changes may be cut-throat competition between groups of people for survival in general on the one hand, and hunger for dominance in the economic, intellectual, educational and political spheres of life, on the other. These changes are tantamount to great expansion of economic organizations, the creation of new institutions to cope with changes leading to occupational and physical mobility horizontally and vertically, the rise of universalistic values and ideologies, and the emergence of a new class structure.

Contemporary societies, such as India, may face a number of problems in the process of change involved in the process of urbanization. Economic and political changes may come in conflict with people's traditional values, attitudes and behaviour. People may like to change but find that change brings with it many unknown problems, tensions and anxieties. Many latent and manifest hostilities towards public and private institutions are generated within urbanizing and industrializing societies and their settlements during the throes of change. It may become possible to arrive at a broad typology of industrially developing societies and their settlements based on type of tensions, resentment and active opposition to new trends. The latter may slow down the industrial-urban process or may push the course of events in a direction of uncertainty and doom. The character of emerging semi-urban realities, as may be found in different developed societies, is likely to spearhead opposite trends and social movements. As Sjoberg states, 'This is especially likely...where traditional values are reinforced by a long standing and deeply entrenched bureaucratic apparatus' (1967b: 222). But outside these traditional societies and settlements, one may find people influenced by egalitarian ethics ingrained in religious ethos, undermining industrial urbanization. The transitional human settlements in traditional societies may remain under the simultaneous impact of various processes of partial change, as suggested by Sjoberg: '(1) the persistence of traditional forms, (2) the revision or modification of traditional forms, (3) the disappearance of traditional forms, and (4) the emergence of new structures' (1967b: 224).

Of course, SUPs are the outcome not only of the influence of old traditional values and behaviour but also of combinations of both aggressive and regressive elements inherent in the human psyche everywhere in the world. The four patterns of social change suggested in the foregoing may not fully explain the complex new trends in urbanization, unless the impact of aggressive elements of change is taken into consideration. Considerable useful analysis of the role of ascriptive traditions in industrializing societies was done in the past, but very little is written on the aggressive import of elements of negative change on urban life and institutions. The contradictory situation—ascriptive traditions in societies like India and universalistic traditions in societies like the US—is such that the semi-urban elements in the first type of societies are tradition-bound, while those in the second type of societies stem from modernized deviant behaviour of individuals and small groups cutting at the root of utilitarian culture in which they would like to live (Gouldner 1970: 61–87).

The centrifugal and centripetal forces characterizing these social movements in both traditional and modern social systems are viewed here at both the micro and macro levels. These tendencies in the changing milieu of a society and in its new type of settlement patterns can be seen in the behaviour of people at all levels of social stratification. If a centripetal force is working at the family level, a centrifugal force may be operating at the larger institutional level, especially in the form of economic and political institutions. These seemingly divergent tendencies among individuals and groups indicate that it is appropriate to discuss the problems of socio-economic change in terms of urbanizing and de-urbanizing societies and their settlements, rather than in terms of the city as such (Reisman 1964: 196).

Hypothesis 3

In every society, irrespective of its stage of traditionalism or modernity, an optimum level of 'stability' is reached such that it leaves room for microscopic areas of differentiation in its social structure and organization, which widens its gaps either due to internal disorders or due to the impact of external change agents, and creates at the macro level vast areas of discontinuities involving many heterogeneous elements, as a result of which SUPs are created in the social system with structural differentiation peculiar to local conditions.

The third hypothesis answers the question as to why the present technological advances make the urbanization process more complex and, at the same time, leave areas of underdevelopment, leading to the emergence of semi-urban situations in the form of human settlements. The historical and instantaneous forces discussed earlier affect every form of contemporary human society, such as, nomadic, primitive food-gathering, pastoral, agricultural and industrial, living in villages, towns and metropolitan areas. These societies may belong either to the little or to the great tradition (Redfield 1960), or be far beyond the traditional, or be at the stage of becoming modern or ultra-modern. But none of them can escape the influences of historical and instantaneous forces affecting the composition of their settlements. These forces are at the back of changes experienced by most of the present-day societies. Thus we come across new structures and organizations both in settlement patterns and the socio-cultural milieu. Increasing numbers of semi-urban small and medium size towns and SUPs in bigger cities come up, where people live in relatively homogeneous groups. In other words, various kinds of SUPs are rapidly increasing in number and size in rural hinterlands and in emerging towns and cities, especially in today's developing countries. In large towns in Africa, for instance, a number of voluntary organizations of indigenous people have come into being for the protection of people coming to cities from diverse ethnic and rural origins, so that they are not forced to get widely scattered and totally absorbed in the alien culture of the city. This subject is ably discussed in various contexts by Gutkind, Mitchel and Bodgan (see Meadows and Mizruchi 1969).

A QUALITATIVE ANALYSIS OF THE GLOBAL PHENOMENON

The revolution in communication and transportation in large parts of the world has led to unprecedented mobility of man and material between the central urban places of politico-economic power and the vast resourceful rural hinterlands. The merging of materialistic and non-materialistic aspects of life is accelerated due to the impact of historical as well as instantaneous forces on human life. In brief, the interpenetration of human beings, raw and finished goods, movement of artifacts, and the diffusion of cultural traits (such as customs, values, ethics, philosophy, law and behavioural patterns) from the rural hinterland to urban centres and *vice versa* have led to a synthesis of urban and non-urban styles of life.

The recent growth of middle class suburbia and its perpetuation in the US is a result of newly married couples wanting to settle in independent homes for reproductive purposes and to keep away from the old tradition-bound society and their parents. But when these young parents become old, the character of suburbia undergoes change because they later tend to lean more towards neo-traditional values and behaviour (Dobriner 1958: 89–143).

The reasons for the emergence of SUPs in developing countries like India are just the reverse. Here, people want to live in the shadow of traditional social structure for the protection of their offsprings and themselves. Whatever be the dominant forces behind the emergence of this third social reality in different countries of the world, and whatever be the prior conditions leading to its emergence, the result is more or less the same.

Whatever be the stage of development of a social milieu and settlement pattern, individuals and groups therein react to external or internal pressures in various ways. Microscopic areas of differentiation arise when individuals and groups decline to accept their ascribed status and roles under changing conditions. When this happens in the traditional spheres of life, old institutions begin to decay, and when this happens in the case of new institutions, it affects their efficient functioning. This leads to the following changes: underdevelopment or partial utilization of economic and technological assets and resources; a lower level of impersonalized secondary sets of relationships; a higher degree of differentiation and discontinuities in status, role and actual behaviour; weak linkages with traditional institutions; low level of participation in politics, civic life and local administration; lack of individual concern for lessening discontinuous linkages for the purpose of integration; and low level of occupational and social mobility. The most important result of these changes is the emergence of disparity in the status, roles and functions of existing institutions and organizations having ascriptive goals. Moreover, the rational outlook disappears from traditional elites due to their inability to respond favourably to changes, change-agents and new reference groups. The continuance of customary laws and usages ultimately results into SUPs of various types.

The concept of social stability, however, is difficult to define because it is related to time and space. A society may attain a higher level of stability when its social structure (network of relationships) and social organization (operation of network of relationships) show the least differentiation of status and roles in the interaction between individuals and groups in day-to-day life. If scattered instances of microscopic differentiation are found due to the exhibition and practice of individuality by some members of a group, their actions may not result in the breakdown of existing patterns of interaction in the society. This means that microscopic differentiation ingrained in the psyche of members of a society may not lead to discontinuities in the functioning of its social organization, until of course they grow to an unmanageable number and attain greater and greater social power.[4] The activities of a small band of people of this kind are generally ignored, but are also likely to be suppressed by legal and quasi-legal

[4] The emergence of microscopic areas of differentiation and change attributed here to individuals and groups in a relatively stable society may be compared with the twin concepts of voluntarism and emergence basic to the theory of Parsons. Although the comparison appears to be rather vague at the present moment, it appears reasonable to quote Devereux Jr (1961: 14), especially with regard to what he says on Parsons' use of the concept of emergence: 'By emergence, he simply means that systems have properties which are not reducible or explainable in terms of the parts which make them up, and that at various levels of organizational complexity ever new orders of systems tend to emerge.' This explains in a crude way, I suppose, the reason why contemporary societies tend to give birth to a variety of SUPs peculiar to local conditions.

measures by power elites, as and when required. The attitude of a nation towards the hippie movement may be to ignore it, but the radical student movement is generally curbed. When any of these movements becomes intense and widespread, it leads to discontinuities in the functioning ability not only of the educational system but also of the entire social organization at different levels (Noble 1982: 251). When such discontinuities occur intermittently, social stability is not likely to be undermined, because then it becomes a part of ongoing social change in the form of acculturation and gradual assimilation.

This phenomenon of change may appear casually in manifest but generally in latent form (Meadows and Mizruchi 1969: 3). In the context of African societies, the problem of stability refers to the gradual settlement of rural folk in towns under non-traditional forms (Mitchel 1969: 470–93). But when these activities take the form of organized and long-drawn out movements, they evoke many heterogeneous social elements which weaken the hold of traditional social structure and organization over the status, roles and functions of individuals and organizations. These changes in various spheres of life may lead to the reorganization of like-minded people into effective pressure groups. Such efforts at differentiated social reorganization may lead to the emergence of SUPs in any society.

RELEVANCE OF THE THEORY TO THE INDIAN SITUATION

After independence, Indian society was subject to many processes of differentiation and discontinuities with regard to the functions of two major institutions—caste and the joint family. This is happening from one end of the spectrum to the other, that is, from the village to the metropolitan city. Being politically open, Indian society is attacked, so to say, by many change-agents in various spheres of life. These change-agents are not necessarily foreigners coming to help and advise the government or non-government agencies in diverse fields of development, but also those from within, that is, the carriers of new scientific knowledge, technology, and universalistic values and attitudes. These include political elites, businessmen, scientists, educationists, technologists, planners, and a variety of trained grassroot functionaries, who are active everywhere in bringing about change by voluntary efforts or through the planning machinery of the government.

There is probably no traditional institution in India with ascribed roles which is outside the sphere of influence of change-agents.

The increasing population and the spread of education in villages have led to the constant migration of younger people from villages to towns and cities. The members of Scheduled Castes and Scheduled Tribes working as farm tenants and labourers also migrate seasonally or permanently to urban areas. Members of these groups tend to live in exclusive clusters of their own mainly in urban areas. Although they may be differentiated and stratified with regard to income, occupation, language, caste and religion, they remain homogeneous with regard to customs, patterns of behaviour, values and living standards.

In the new middle class colonies or in the old pockets of towns and cities (like the *katras* in Old Delhi or the *poles* in Ahmedabad), caste affiliations not only persist but get politicized and reorganized to meet new threats and challenges faced by their members (Trivedi 1980). In the political sphere they are fragmented and do not behave as monolithic socio-cultural solidarities. The double standard that appears in the economic and political behaviour of caste groups is relatively less significant in rural compared to urban SUPs (Trivedi 1971). Caste solidarities have suffered a setback in urban areas due to the spatial dispersal of caste members as well as the differentiation of statuses and roles assigned to them in different types of organizations and institutions. The present situation with regard to caste and other ethnic group relationships is getting more and more complex with unprecedented changes in economic and other spheres of life. The government policies of reservation and promotion for Backward Classes, technological advancement, and ever-increasing communication networks are at the root of the complexity. There are now far more statuses and roles than at any time in the past, leading to the formation of class consciousness in urbanizing areas, impinging on the solidarities of the ethnic group, sub-caste, traditional joint family and the multi-caste village community.

The type of wholeness that Durkheim saw in the Western city, which maintained organic balance in spite of social change within it, is not observable in larger towns and cities in India today. The network of interdependence of various groups has now fallen into partial confusion and tries for re-adjustment through caste organizations. The cross-currents of incipient class, caste and interest groups in towns and cities are sensitive to current political and economic developments. This sensitivity takes destructive forms under extreme conditions of industrial strikes and *gheraos* (to surround a person or persons in authority until one's demand is met), confrontation with the civic authority, conflicts between ethnic and religious groups, and multi-party political upheavals (Trivedi 1969b). After more than forty years of independence, the organic solidarity in towns and cities in India is disturbed more due to conflicts between ethnic groups and diverse communities than other reasons.

The SUPs are not immune to such extreme conditions of temporary disorganization because the people there have low living standards and a limited impact of urbanization. In relation to ethnic variety and number, however, they have not experienced large-scale class conflicts, except in the form of limited student trouble of a temporary nature. The caste and familial traditions, far removed from the control of the integrative forces of class and community ethos, continue to dominate. In general, changes in work situations and occupational interests of the people in SUPs do not bring about corresponding changes in the structural relationships of the people. Their social systems continue to operate under forces converging in favour of caste structure than of class and the urban community in a wider sense. Here the primary relationships are not at stake as much as the secondary ones. Secondary relationships are found only in work situations (like business establishments, industries, cooperatives, clubs, educational institutions and agencies for social welfare). The family life of individuals in SUPs provides greater scope for the mingling of individuals and families belonging to traditional social groups. As a result, the differentiation of newly acquired status and roles in economic and occupational fields is considerably subordinated to traditional group solidarity. Had this not been so, it would have enhanced the formation of diverse class structures in India (Anderson and Ishwaran 1965: 87–88).

My study entitled 'The Process of Urbanization and Social Change on the Fringe of Ahmedabad City' (Trivedi 1981) provides a striking illustration of the proliferation of semi-urban development of the type suggested in the foregoing. It is a study of 130 housing societies built more or less by cooperative efforts of households living in a broadly heterogeneous milieu. After 1981, many more housing societies of high castes, artisans, Scheduled Castes, Muslims and Backward Classes working in banks, government departments, educational institutions, and so on, have been added. These housing societies have been built on plots of village land put to non-agricultural use under the Town Planning Schemes approved by the Ahmedabad Urban Development Authority. Most of the members of the households here lead tradition-bound ascriptive lives.

AN APPRAISAL OF OLDER THEORIES OF URBANIZATION

In the light of the foregoing analysis, it will be worthwhile to make an appraisal of various schools of thought related to urbanization. Such an appraisal is presented here in a sketchy way (for an extended discussion see Trivedi 1975). In the present exercise of attaining a higher level of abstraction, it is neither necessary nor feasible to cover all the findings of all schools of thought discussed extensively in that book.

Sjoberg (1967a: 157–89) includes Park and Burgess (as also their colleagues and students), Louis Wirth and Robert Redfield in the growth-oriented school which propagates the materialistic approach to the study of urbanization. This approach covers the concept of the city with respect to size, density and heterogeneity, and the processes and patterns of transition from the pre-industrial, agrarian and feudal order to the industrial, urban and capitalist order.

Sjoberg groups the works of Whyte, Gans and Oscar Lewis under another school. Though they are essentially sociologists, their works are compared with noted ecologists, particularly because all of them had emphasized the importance of relatively homogeneous social areas in a city. In fact, they are entitled to be grouped under the sub-social area school which challenges the urbanization school of ecologists for the contention that the city is a dependent variable—dependent on the behaviour of the people.

The real distinction between these two schools appears to be that although the city is looked upon as a dependent variable with regard to its materialistic aspects of life, it is not really so, because a large part of the social or natural areas within the limits of a city show non-materialistic characteristics independent of the whole. As indicated earlier, certain natural and social areas in the city are identified by me as urban SUPs. In a general sense, however, they represent either concrete or symbolized non-materialistic aspects of city life.

The ecological complex school of Duncan and Schnore and the sustenance school of Gibbs and Martin highlight only the materialistic aspects of city life. They lay emphasis on the complex of environment, population, social organization and technology on the one hand, and sustenance-cum-behavioural activities on the other. They exclude people's value orientations as an essential element either in aggregate social terms or in relation to values attributed to the division of labour. Here also the implication of superiority of the materialistic over the non-materialistic aspects is the main drawback. As Sjoberg puts it, 'these modern ecologists have confused the study of a substantive problem...with a particular theoretical perspective' (1967a: 168). The purview of the ecological complex school no doubt goes beyond the confines of the city to the region, but the idea of the city as conceived by them is probably too sacred to be divisible on materialistic and non-materialistic levels of abstraction.

The economic school led by Colin Clark speaks of the increase in the 'scale of society' in its social interaction and dependence as it moves from the primary to the tertiary means of production. The structural indicators of the scales and levels of production are '(1) changes in the distribution of skills, (2) changes in the structure of productive activity, and (3) changes in the composition of the population' (Sjoberg 1967a: 168–69).

The social area analysis school of Shevky and Bell largely collaborates with manipulating census tract data and fails to explicate its theory. Here also there is too much spider-work on economic variables, which makes it difficult to attain any kind of higher abstraction beyond the concept of the city as an economic organization.

The weakness of the environmental school of Mumford lies in the belief that modern man cannot control his destiny in the midst of uncontrollable environmental conditions, unless he strikes a balance between nature and the city (such as Athens) as a suitable artifact. This school holds a circumspect and traditionalist view of an ideal city type (Sjoberg 1967a: 169).

The imperfection of the technological school is that it emphasizes the impact of technology on the spatial ordering of new elements within the city and *vice versa*. Moreover, it does not define it 'in such a manner as to make it operationally researchable, yet theoretically meaningful' (Sjoberg 1967a: 171). Besides, it is as deeply inward looking as the economic school, focusing its attention mainly on the materialistic aspects of the city.

The value orientation school becomes significant when one compares ancient cities in highly divergent cultures and agrees with the viewpoint that the ecological order, physical planning and social organization of a city are markedly influenced by cultural and particularly religious values of a people. This phenomenon may be an overt manifestation of the values internalized by the people. In complex societies, however, it is difficult to assess strong connections between the value system and the action pattern of individuals and groups. For scientific purposes, however, the significance of this frame is undermined because of difficulties in distinguishing between values, ideas, beliefs, and so on (Sjoberg 1967a: 179). Besides, there is the danger of overemphasizing a particular system of values as being conducive to the city's growth rather than an objective understanding of the impact of values in general on contemporary urban society.

The social power school stresses the influence of the political power of urban elites in shaping the city (Sjoberg 1967a: 174–78). Its area of concern covers the influence of social, religious and economic elites coming from diverse groups living in a city. Government institutions and voluntary organizations also influence the spatial and land-use patterns in a city. The influence of political power on the city operates in three different spheres—local interests, national goals and policies, and international events arising out of wars, natural catastrophes, and so on. In American cities, the land-use pattern is a product of the bargaining power of business elites and the adjustment of competing interest groups. In some African countries the policy of apartheid enforces the spatial distribution of the native population away from central zones occupied by White rulers. Likewise, the struggle for dominance of the conquered cities by warring nations leads to different types of organizational and spatial structures of cities, as found in the emergence of two Berlins after World War II.

An overall view of the various schools indicates that the impact of social power on cities is most important. Its patent examples are the growth of suburbia and slums of different types in the American city, and the emergence of SUPs in India and other developing countries.[5] As rightly pointed out by Sjoberg (1967a: 178–79), the theoretical implications of urban community power for urban settlements deserve

[5] My comments on various schools of thought are based on what Sjoberg has said about the contributions of various scholars on the subject, and the extent to which I have been able to understand him. I have not studied the original works of these authors, except of those included in the references that follow. The responsibility of misinterpretation of facts with regard to these or any other references in the article is entirely mine.

to be further explored. This paper aims at the same goal, but emphasizes that the impact of ethnic community power leads to the concentration of populations in specific areas of a city in underdeveloped countries, and gives rise to a third social reality, in the form of SUPs, in the fast urbanizing world of today. In brief, the community power hidden in places of population concentration and behavioural aspects renders the materialistic viewpoint insignificant for a general theory of urbanization.

NEW TRENDS IN THE THEORY OF URBANIZATION

In an overview of urbanization phenomenon, Philip Hauser (see Hauser and Schnore 1967: 41) has rightly concluded that most of the failures in understanding the nature and content of urbanization lies in our inability to clearly distinguish between its dependent and independent variables. The distinction between the two kinds of variables is usually drawn without a clear perception of the total reality. No doubt there are overlaps between the two. Granting that such dilemmas are difficult to resolve by framing watertight compartments, the reasonable thing to do is to identify a variety of mixed borderline settlements, such as suburbia and SUPs appearing in various forms (for instance, symbolic, physical or value oriented social realities). This mechanism lends clarity to the perception that the concept of the city as a single monolithic whole will have to be discarded. It is also reasonable to accept that the semi-urban reality with behavioural over-tones exists in the social environment within as well as beyond the city limits. We should see the new semi-urban reality also in its physical dimension with accompanying materialistic components.

It is questionable whether the multidisciplinary approach suggested by Hauser will be of any help in attaining a fuller understanding of urbanization, because it is difficult to fuse together the ideological orientations and methodological preoccupations of scholars in each discipline that could be helpful in gaining the required level of abstraction. This is the shortcoming of interdisciplinary teamwork aiming to theorize on a highly complex subject like this.

To undertake comparative and historical studies, as suggested by almost all authors on the subject, is essential, but these need not take pre-industrial cities as the starting point with the hope of building a theory in terms of the evolutionary development of urban phenomenon. Studies of overall change in industrialized as well as developed and developing societies alone can be useful in such an exercise. The consequences of technological development and the related trends of urbanization have qualitative similarities in most societies today.

In the new attempts at theorizing on the city in particular and urbanization in general, the names of Reisman (1964), Sjoberg (1967a, 1967b) and McGee (1971) stand out prominently. Reisman has provided a useful typology for a comparative description of urbanization on a global scale. The strategy is to compare cities at different stages of development in a wide range of societies, classified on the basis of urbanization variables, namely, the growth of urbanism and industrialism, the emergence of the middle class, and nationalism. Such analysis would certainly help compare the institutional structures, attitudes, politico-economic values and ecological features of cities. Reisman believes that it has a theoretical potential and that 'it should appear in the differences in the cities at different points in the development process' (1964: 236). He has taken pains to understand the character and processes of urbanization in different societies of the world in two respects—first, at the material development stage and, second, at the concomitant stage of changes in the growth of population in towns and cities. The weakness of this theoretical proposition is that it continues to equate the terms 'city' and 'urban' in the traditional fashion, with emphasis on the materialistic features of urbanization.

In a somewhat reverse process of reasoning, Sjoberg has rejected 'a "city-limit" kind of sociology in favour of a study of the city in relation to the broader socio-economic context' (1967b: 252). He also focuses attention primarily on the industrial city just as Reisman does, but tries to bring out a general outline of the growth of urbanization on the basis of the pre-industrial city's social structure. He, however, admits that it is untenable to assume that urban social systems are neatly and consistently arranged realities, and that structural contradictions within industrial cities and societies persist. The existence of contradictory functional requirements of a city, according to Sjoberg, are associated with contradictory structures within an industrial city and society, and are reflected in the findings of social scientists who try to delineate significant areas of disagreement in the perception of urban reality. The concept of SUP is proposed to resolve these contradictions and to give a clearer perception of urbanization and urbanism that are ever changing within the range of rural-urban polar realities, both interlocking and reshaping themselves perpetually in interaction with one another (*cf.* Mangin 1970: xiv–xix).

The conclusions derived by Sjoberg (1967b: 252) in his comparative analysis of cities in industrial and industrializing societies with technological base as the key variable do not seem to have untainted clarity. He says that although there is significant agreement on the nature of urbanization, there is confusion of apparently contradictory data presented by different scholars. The main cause of confusion is that the city is taken as an indivisible entity, although it comprises many divergent social realities identifiable either in an organic economic sense or in a non-organic cultural sense, and also in behavioural relationships of people within the urban system as a whole. It is probable that if seemingly contradictory data are noted in the light of the concept of SUP, it may help to resolve the contradictions. As behavioural aspects of social structures are not neatly and consistently arranged, valid comparisons cannot be made without resolving contradictions in the perception of urban social structures.

The problem of division of labour is not necessarily confined to the so-called industrial and pre-industrial social systems, when we carefully examine as to how it worked in primitive societies, either as a divisive or integrative force under different conditions. Whether a society enjoys mechanical or organic solidarity, the division of labour, especially in the service sector of the economy, is bound to give rise to a set of mediators (such as priests, princes, managers, courtiers, negotiators, compromisers and politicians). The question whether these mediators are whole-time or part-time operators needs to be understood. If they are whole-time workers and constitute a large number, they may organize themselves in the form of a class in the mosaic of complex division of labour and social relationships of which they are the products. But if they are marginal or part-time workers, they may not form a class as each of them may perform a substantive ascriptive role in their respective traditional system of social stratification.

In industrialized societies (such as the US) or developing societies (such as India), the urban population inside or outside the limits of city government have both full-time and part-time mediators involved in the economic, social, religious and other aspects of the division of labour. These mediators are active in different social settings in highly differentiated and sophisticated spheres of urban life. Likewise, they are active in the simpler and relatively homogeneous settings of caste, religion and ethnic groups in SUPs. If distinctions between the roles and functions of mediators and other workers are not properly understood with regard to their significance at the time of studying urban problems, the findings of specific studies are likely to give contradictory results. Therefore, to improve the quality of understanding of the division of labour and the role of mediators in urban social areas, a study of their values, attitudes and behaviour in the light of particularistic and universalistic orientations is necessary.

McGee's (1971: 55–58) classification of cities into 'stable' and 'migrant' has been an important step in evolving a typology related to pre-industrial, colonial and industrial and social dimensions on one side, and peasant characteristics on the other. The latter are attributes of people migrating from rural areas to different types of cities and urban agglomerations. Here also, the theory takes into cognizance basically one-way traffic of people from the hinterland to urban centres. The behavioural process of penetration of urban elements into rural settlements is not spelt out. This means that the terms 'city' and 'urban' are looked upon as synonyms, which is the basic pitfall from which his theory cannot be extricated. The emphasis seems to be on the ruralization of cities alone rather than on the urbanization of rural areas as well.

In contrast to these major attempts at theorizing, the framework I have presented in the Indian context (Trivedi 1971) has kept in mind the typology of SUPs, which synthesizes the rural-urban differential in the historical and spatial perspective and recognizes thereby a two-way traffic between rural and urban components. It gives due recognition to the on-going processes of interpenetration and accommodation in a variety of ways, in multiple combinations of adjustment and co-existence of ethnic groups with varied backgrounds. The morphology of rural and urban SUPs should also help in classifying Indian cities and towns on the basis of preponderance of SUP types as well as combinations of sub-types under each major system of classification. It is, however, necessary that every country arrives at its own typology of rural and urban SUPs, on the same trichotomous conceptual model, in the light of local variations.

A CRITICAL APPRAISAL OF SELECT RECENT LITERATURE

Most writers on urbanization today have been reluctant and cautious about propagating a general theory of urbanization and social change.

Every country has unique historical, political, geographic, economic and socio-cultural antecedents that inhibit such an effort. Behaviour is also shaped by values and related attitudes, of which the overall result remains specificity-oriented action. The act of barring specificity variables from analysis is difficult, but not impossible if a higher level of abstraction is reached (as in the case of the SUP model evolved in this article). I now make an appraisal of some recently published literature to justify this model. Urbanism and the urbanization process are highly complex and will continue to perplex and dodge clear-cut identification, especially because they involve the interplay of two core dimensions of the material and non-material aspects of human existence.

Let us begin with Africa. The rich data presented in *The African City* (O'Connor 1983) are fascinating. The author has considered the city as a monolithic concept and compared the urbanization process in most parts of the world. He has identified six typologies in relation to African cities which developed certain common features in the midst of ethnic diversity, intense rural-urban movement, and poverty in material development over a long period of time. His ambivalent analysis is a clear admission of microscopic areas of differentiation in rural-urban relationships that persist in most parts of the world. He says:

> one conclusion that has emerged from writing this book is that 'urban' and 'rural' are categories that should be used with great caution, and perhaps rather sparingly in tropical Africa. In much of the Middle East, in southern and eastern Europe and in Latin America one can distinguish fairly clearly not only rural/urban places but also rural/urban people. Such a distinction is less clear-cut in the United States or north-west Europe, and in a very different cultural and economic context it is equally blurred in tropical Africa (1983: 306–7).

Whether we take geographical features or social change as the wider cause of urbanization, we have to accept that it is mainly a mental construct with behavioural undertones.

Guglar and Flanagan (1978) provide a perceptual frame of three types of change—historical, biographic and situational—that form the dynamic essence of this general theory of urbanization. 'How fast do the three types of change that we have distinguished proceed? Situational change is abrupt by definition. Historical and biographic change tends to occur more slowly. However, societies do experience abrupt change of great magnitude' (ibid.: 117). Although change is a force to reckon with in the study of urban theory, the forces of continuity also work to shape urban life-style. 'A traditionalist outlook on the part of an immigrant encourages incapsulation in a close-knit network of like-minded associates who enforce conformity with the pattern prescribed by the group' (ibid.: 116). But cultural background is not the only binding force behind social togetherness. The social density characterizing love for village ethos is equally important for the tendency toward growth of the semi-urban type of settlements.

Hanna and Hanna (1981) and Peach, Robinson and Smith (1981) indicate how residential segregation is a persistant phenomenon among the natives themselves in Africa. There is accommodation but no assimilation between ethnic groups migrating to urban areas, where the rural type of SUPs can be identified in profuse numbers. 'The city continues as it was envisaged by the Chicago ecologists—a pattern of territoriality maintained by social interaction' (Hanna and Hanna 1981: 21). However, the scope of change and its direction are issues of great significance. As Eisenstadt puts it:

> the direction and scope of change are not random but dependent...on the nature of the system generating the change, on its values, norms and the organisations, on the various internal forces operating within it and the external forces to which it is specially sensitive because of its systemic properties (1964: 247).

This echoes what has been expressed in the three hypotheses providing a viable model and a general theory of urbanization in this paper.

In a perceptive analysis of the Chandigarh experiment in urban planning in India, Sarin (1982) has boldly brought out a profile of the latent social situation of the underprivileged, semi-skilled people whose life-style is manifested in the form of informal occupations, and squatter and improvized non-planned settlements. The people of this service sector are of rural origin and retain old kinship linkages which give them semi-urban characteristics. Concluding her fascinating and down-to-earth presentation of the city planning, Sarin writes:

> Given the scarcity of skills and resources among the masses of the working poor, clearly none of these aspects can be dealt with at the individual level. The only alternative lies in creating structures

> which increase the socialization of labour, particularly among those in the ever-growing tertiary or 'informal' sector of employment with the clear objective of enabling them to alter the structure of production, distribution and decision-making (1982: 254).

It seems that in a country like India, city planning should make ample room for semi-urban settlements that can absorb cheap labour of the rural working immigrants.

Though urbanization is nebulous and ever-changing in all the countries of the world, this is not true of China. Until recently, Chinese towns did not have shanty settlements. As rural migrations to cities were strictly controlled, the urban poor were almost absent. The general profile of the metropolises was gentle and village-like social life. The hinterlands were largely rural, but the cities and counties exhibited sprawling semi-urban settlement patterns. In a critical assessment Kerkby (1985: 253–54) states:

> China spares many of the difficulties of development faced by the so-called developing nations. Where China's experience of urbanization is unique, it is not because the distinction between the rural and urban worlds has been overcome. China's claim to a special place in the annals of urbanization must rest on two pillars of political and social policy peculiar to its culture and polity. ... One of their abilities is to control overurbanization by the systematic prevention of an exodus of rural migrants. The second is the achievement of substantial industrial growth without the urban misery. This does not mean that the Chinese population has exclusive divisions of only urban and poor population groups. On the contrary, a large section of Chinese society is semi-urban, especially in terms of values held by the masses under the pressure of political doctrine.

Hall (1977) deals with urban-regional change in *Europe–2000*, which is related to technological developments and changes in the economy, society, life-style and values with expected changes in the immediate future and an impact on patterns of urban-regional society. He reports a kind of reverse trend in urbanization akin to the emergence of rural semi-urban settlements, which is also comparable to what is happening in America. According to Hall, the prospect of change in Europe is one of a partial return to spatial patterns of the agrarian Middle Ages.

With the uniformly available advanced technology infrastructure, a high degree of residential dispersal will be possible. Many people will be free to express a preference for isolated farms, villages and small towns. Negative externalities of crowding in the larger metropolitan areas may mean that for those workers free to choose residence, the older locations are preferred, which are virtually rural settlements in regions unaffected by the first industrial revolution.

Though French historiographers have not produced models on urbanization they have indulged recently in theoretical reflections (Bedarida 1983: 402–6). First, the concept of function invariably leads them to compare the town with an organism where the heart, lungs and arteries operate in unison. It is admitted as illegitimate comparative analysis in terms of organism. Second, the relationships between classes and their fractions condense and sharpen the economic contradictions due to the occupation of the valuable urban span by the dominant class even under Marxist regimes. That is why Marxist thoughts on the theory of urbanization get dissolved into the social question in totality. Third, new models of social and family behaviour imposing norms on groups of the depressed class makes the town a place of power, both social and economic. But the highest power is that of the development rationality where individuals no longer exist meaningfully as normalized beings to the exclusion of emotions, desires and personal relations. Fourth, urban semiology based on anthropology and structural linguistics has tried to bring a new dimension to the study of the history of towns and to that of urbanism by opening up new fields of research on town space and the operative concept of town planning (ibid.). Furthermore, 'A town, indeed a shifting totality in a perpetual state of becoming, can be read in many different ways. ... Each group reads its town after its own fashion' (ibid.). Finally, there is the approach based on systems theory that is at the crossroads of urban history and social sciences. To put it clearly, the system organizes its own regulation through the strategies of individuals and groups, reproducing itself by means of its territorializations and mutations. The three hypotheses presented in this paper covers genuine theoretical points of view in the generalization of the phenomenon.

Gappert and Knight (1982) provide a fascinating picture of the future of American cities in the twenty-first century. The realignment of heavy industry in the United States has resulted in unprecedented changes in the viability of neighbourhoods.

> Since 1970, a process has been occurring that Berry (1976) refers to as 'counter urbanization', in which metropolitan areas, especially in the north-central region, have been losing population to non-metropolitan areas. This has resulted in a depopulation of large central cities. Downs observes that from 1970 to 1975, 97 of the 153 largest cities lost population—compared to only 56 that lost population in the 1960s (Gappert and Knight 1982: 104).

In view of this, the intervention programme to control change was identified as the creation of moderate income neighbourhoods where preferences and motivations for living would be more differentiated, as in the case of urban SUPs.

While writing on developing a comparative sociological theory of urbanization, Thorns (1982) emphasizes the form of the productive system and the dominant values of the society. He assumes the city as the monolithic centre of urbanization and states that the fabric of city-producing urban patterns are related to the concerns of factorial ecologists and social area analysts. Capitalist societies, according to him, are divided into different areas on the basis of socio-economic status, familism and ethnicity.

> These differences in the pattern and magnitude are not fixed but change over time. The establishment of these changes between cities does not answer the question as to why they exist, how they come into being, nor why they differ. These then are central questions for a comparative urban theory to engage (1982: 39).

The questions raised by Thorns are value oriented. He lays importance on the dominant values of society but does not relate them to the attitude and behaviour of the rural people who contribute their might as elements of the working force in the development and growth of cities. The development of the city in Australia and New Zealand taken as illustrations of land market operations may be related to the changing city without reference to the sections of native outsiders visiting or settling in the cities built on the Western model.

In *Urban Politics: A Sociological Interpretation*, Saunders writes,

> Both the American concern with community power and the British interest in urban managerialism reflected a growing dissatisfaction

> with ecological theories which sought to portray as 'natural' what was clearly the result of human action and decision making, and to discuss the individual actor as insignificant when he was in fact central to the urban problematics (1983: 138–39).

Here two fallacies seem to be involved. First, the scholars of the ecological school of Robert Park used the term 'social areas' as if they were dense pockets of city settlements, indicating the tendency of various rural immigrants to live in social homogeneity. These were natural areas inhabited by people known to them intimately and harbouring similar values, attitudes and behaviour due to kinship or community feeling, and sometimes coming from the same rural areas. As immigrant settlers, their primary concern was not to take hold of power but to keep their social norms and customs intact and sub-cultural identity homogeneous. Those who were outward looking and interested in change and power politics were the elites who stood above the ground of the people living in natural areas. They were a handful of power elites—the economically, educationally and politically dominant but numerically small and socially heterogenous. They dominated urban planning and the distribution of scarcer resources to safeguard their interests first, and not to benefit deprived ethnic groups. They are exploiters first and benefactors later. It is true that the emphasis of scholars' concern in most countries shifted from biotic struggle to the analysis of power and its distribution. But the allocation of resources for zoning and micro-level development remained inequitable. The rural immigrant groups had to remain possibly in the state of semi-urban pockets in the natural areas and ghettos of towns and cities.

Eisenstadt and Shachar have achieved a landmark in presenting a macro view of theories of urbanization in the context of almost all shades of culture. In the epilogue of their valuable book *Society, Culture and Urbanisation* (1987), they have employed the comparative method to analyze the phenomena in nine civilizations of diverse nature. The review of numerous theories have been analyzed by fusing innovative analytical approaches. This has been attained to help understand the evolution of cities over time, space and cultural background. The most noteworthy stance they have taken on universalizing theoretical concern is expressed as follows. 'In analyzing the social actors, forces and processes that in various civilizational contexts influence the development of aspects of urban phenomena, we have subsumed these forces under two headings: "concentration" and "centrality". Needless to say, this distinction is an analytical one and not a description of concrete social phenomena' (1987: 358). That the combination of demographic, technological and socio-economic forces and institutions has the capacity to shape different aspects of the urban phenomena as envisaged in my hypotheses and model of semi-urban pockets is echoed by Eisenstadt and Shachar:

> whatever the differences between the transformations in post-industrial societies and those in the so-called third world, the common core of the forces of modernity of communication and information technology in general and of the spread of new types of political ideologies in particular has greatly changed the nature and structure of the instrumental and symbolic institutional aspects of urban-rural relations, as well as of the distinction between centre and periphery (1987: 362).

CONCLUDING REMARKS

A general theory of urbanization has to be a general theory of social change. Urbanization in developing societies and de-urbanization in developed societies appear to be the principal sources of social change. The channels of rapid and widespread communication and new technology tend to disrupt, reorganize and stabilize the contemporary social systems in the form of agglomerations and pockets. In this process of change all over the world, a trichotomous classification of social systems can be perceived as rural, urban and semi-urban. The concept of SUP, representing a combination of materialistic, non-materialistic and symbolic aspects of the urban phenomenon would lead us to arrive at the following qualitative classification of SUPs for study in future: (*a*) SUPs with a higher content of materialistic aspects of urbanization, (*b*) SUPs with a higher content of non-materialistic aspects of urbanization, and (*c*) SUPs representing symbolic actions and forms of behaviour of individuals and groups that do not conform either to the urban value system or to that way of life.

The semi-urban reality exists at the micro, meso and macro levels of the social situation in many parts of the world today. This is the reason for the necessity to look upon semi-urban reality not only in the form of SUPs at the micro level, but also in the form of semi-urban proliferation at the meso level and semi-urban periphery at the macro level of the changing social situations in the world today.

I would like to reiterate that the main thought behind evolving the concept of a semi-urban pocket is to pinpoint the significance of a social reality that is neither urban nor rural. The focus of this theory is mainly the idea of 'semi-urban' and not necessarily that of 'pocket', suggesting concentration and centrality related to human behaviour and development. That is why one can justify the idea of semi-urban periphery, proliferation, pocket, and so on, irrespective of the nature, size and distribution of the population from one pole of the socio-cultural situation to the other and *vice-versa*. The term 'pocket' as a metaphor can easily be replaced by 'situation' or any other suitable term relevant to a society. The idea of a semi-urban situation seems to be the most enduring everywhere today, especially in the developed countries. However, semi-urban situations in human behaviour have a greater propensity to get converted into types of pocket settlements.

REFERENCES

ANDERSON, NELS and ISHWARAN, K. 1965. *Urban Sociology.* Bombay: Asia.

BEDARIDA, FRANCOIS. 1983. 'The French Approach to Urban History'. In D. Fraser and A. Sutcliffe (Eds.), *The Pursuit of Urban History.* London: Arnold.

BELL, WENDELL. 1969. 'Urban Neighbourhoods and Individual Behaviours'. In P. Meadows and E. Mizruchi (Eds.), *Urbanism, Urbanization and Change: Comparitive Perspectives.* Massachusetts: Addison-Wesley.

BERRY, B. J. L. 1976. *Urbanisation and Counter Urbanisation. Sage Urban Affairs Annual Reviews No. 11.* Beverly Hills: Sage.

BOGDAN, R. 1969. 'Youth Clubs in a West African City'. In P. Meadows and E. Mizruchi (Eds.), *Urbanism, Urbanization and Change: Comparative Perspectives.* Massachusetts: Addison-Wesley.

DEVEREUX JR, EDWARD C. 1961. 'Parsons' Sociological Theory', In Max Black (Ed.), *Social Theories of Talcott Parsons.* Englewood Cliffs: Prentice-Hall.

DOBRINER, WILLIAM M. (Ed.). 1958. *The Suburban Community.* New York: Putnam.

DOXIADIS, C. A. 1967. *Ekistics and Sociology.* Athens: Doxiadis.

EISENSTADT, S. N. 1964. *Social Change and Modernisation in African Societies South of Sahara.* Jerusalem: The Hebrew University.

EISENTADT, S. N. and SHACHAR, A. 1987. *Society, Culture and Urbanisation.* New Delhi: Sage.

FAIRCHILD, H. P. (Ed.). 1955. *Dictionary of Sociology.* Iowa: Adams.

FRASER, D. and SUTCLIFFE, ANTHONY (Eds.). 1983. *The Pursuit of Urban History.* London: Arnold.

GANS, HERBERT J. 1962. *Urban Villagers.* New York: Free Press.

GAPPERT, GARY and KNIGHT, RICHARD V. 1982. *Cities in the 21st Century.* New Delhi: Sage.

GIBBS, JACK P. (Ed.). 1961. *Urban Research Methods*. New York: Van Nostrand.

GIBBS, J. P. and MARTIN, W. T. 1969. 'Urbanisation, Technology, and Division of Labour: International Patterns'. In P. Meadows and E. Mizruchi (Eds.), *Urbanism, Urbanization and Change: Comparative Perspectives*. Massachusetts: Addison-Wesley.

GOLEMBIEWSKI, ROBERT T., BULLOCK III, CHARLES S. and ROGERS Jr, HARRELL R. (Eds.). 1970. *The New Politics*. New York: McGraw Hill.

GOULDNER, ALVIN W. 1971. *The Coming Crisis of Western Sociology*. London: Heinemann.

GUGLAR, JOSEF and FLANAGAN, WILLIAM G. 1978. *Urbanization and Social Change in West Africa*. Cambridge: Cambridge University Press.

GUTKIND, P. C. W. 1969. 'African Urban Family Life and the Urban System'. In P. Meadows and E. Mizruchi (Eds.), *Urbanism, Urbanisation and Change: Comparative Perspectives*. Massachusetts: Addison-Wesley.

HALL, P. 1977. *Europe 2000*. London: Duckworth.

HANNA, W. J. and HANNA, J. K. 1981. *Urban Dynamics in Black Africa*. New York: Aldine.

HAUSER, PHILIP M. and SCHNORE, LEO F. (Eds.). 1966. *The Study of Urbanisation*. New York: John Wiley.

KERKBY, R. J. R. 1985. *Urbanisation in China*. London: Croom.

MANGIN, WILLIAM (Ed.). 1970. *Peasants in Cities*. Boston: Houghton-Mifflin.

MCGEE, T. J. 1971. *The Urbanization Process in the Third World*. London: C. Bell.

MEADOWS, P. 1969. 'The City, Technology and History'. In P. Meadows and E. Mizruchi (Eds.), *Urbanism, Urbanisation and Change: Comparative Perspectives*. Massachusetts: Addison-Wesley.

MEADOWS, PAUL, and MIZRUCHI, EPHRAIM (Eds.). 1969. *Urbanism, Urbanisation and Change: Comparative Perspectives*. Massachusetts: Addison-Wesley.

MITCHEL, J. C. 1969. 'Urbanization, Detribalization, Stabilization and Urban Commitment in Southern Africa'. In P. Meadows and E. Mizruchi (Eds.), *Urbanism, Urbanisation and Change: Comparative Perspectives*. Massachusetts: Addison-Wesley.

MORRIS, R. N. 1968. *Urban Sociology*. London: Allen and Unwin.

MUMFORD, LEWIS. 1961. *The City in History*. London: Becker and Warburg.

MURVAR, V. 1969. 'Occidental versus Oriental City'. In P. Meadows and E. Mizruchi (Eds.), *Urbanism, Urbanisation and Change: Comparative Perspectives*. Massachusetts: Addison-Wesley.

NOBLE, T. 1982. 'Recent Sociology, Capitalism and the Coming Crisis', *British Journal of Sociology*, 23: 238–53.

O'CONNOR, ANTHONY. 1983. *The African City*. London: Huchinson.

ORLEANS, P. 1969. 'Robert Park and Social Area Analysis'. In P. Meadows and E. Mizruchi (Eds.), *Urbanism, Urbanisation and Change: Comparative Perspectives*. Massachusetts: Addison-Wesley.

PARK, ROBERT E. 1952. *Human Communities*. New York: Free Press.

PEACH, CERIE, ROBINSON, V. and SMITH, S. 1981. *Ethnic Segregation in Cities*. London: Croom.

POPENOE, D. 1969. 'On the Meaning of Urban in Urban Studies'. In P. Meadows and E. Mizruchi (Eds.), *Urbanism, Urbanisation and Changes Comparative Perspectives*. Massachusetts: Addison-Wesley.

REDFIELD, ROBERT. 1960. *The Little Community: Peasant Society and Culture*. Chicago: Phoenix.

REISSMAN, D. 1958. 'The Suburban Sadness'. In W. M. Dobriner (Ed.), *The Suburban Community*. New York: Putnam.

REISMAN, LEONARD. 1964. *The Urban Process.* London: Collier Macmillan.

REISS JR, ALBERT. 1964. *Louis Wirth: On Cities and Social Life*. Chicago: University of Chicago Press.

RUDOLPH, LLOYD I. and RUDOLPH, SUSANNE HOEBER. 1967. *The Modernity of Tradition.* Bombay: Orient Longman.

SARIN, MADHU. 1982. *Urban Planning in the Third World: The Chandigarh Experience*. London: Mansell.

SAUNDERS, PETER. 1983. *Urban Politics: A Sociological Interpretation.* London: Hutchinson.

SINGER, MILTON. 1954. 'The Cultural Role of Cities', *Economic Development and Cultural Change*, 3: 53–73.

SJOBERG, GIDEON. 1967a. 'Theory and Research in Urban Sociology'. In P. M. Hauser and L. F. Schnore (Eds.), *The Study of Urbanisation.* New York: John Wiley.

——. 1967b. 'Cities in Developing and in Industrial Societies', In P. M. Hauser and L. F. Schnore (Eds.), *The Study of Urbanisation.* New York: John Wiley.

TARVER, JAMES DE. 1958. 'Suburbanization of Retail Trade in the Standard Metropolitan Areas in the United States, 1948–1954'. In W. M. Dobriner (Ed.), *The Suburban Community*. New York: Putnam.

THORNS, DAVID C. 1982. 'Problems in the Development of a Comparative Sociological Theory of Urbanisation'. In Ray Forest, Jeff Herderson and Peterson Williams (Eds.), *Urban Political Economy, and Social Theory*. Hampshire: Gower.

TRIVEDI, HARSHAD R. 1969a. 'The Semi-Urban Pocket as a Concept and Reality in India', *Human Organization*, 28: 72–77.

——. 1969b. 'Productivity and Social Factors', *Productivity*, 10: 291–94.

——. 1971. 'A Theory of Urbanization in India', *Indian Journal of Social Work*, 32: 267–77.

——. 1975. *Urbanization and Macrosocial Change*. Allahabad: Chugh.

——. 1976. *Urbanism: A New Outlook*. Delhi: Atma Ram.

——. 1980. *Housing and Community in Old Delhi*. Delhi: Atma Ram.

——. 1981. 'The Process of Urbanization and Social Change on the Fringe of Ahmedabad City. Ahmedabad: Institute of Cultural and Urban Anthropology. Mimeo.

WEBER, MAX. 1958. *The City*. Glencoe: Free Press.

Notes on Contributors

F. G. BAILEY, Professor of Anthropology, University of California at San Diego. Author of *Caste and the Economic Frontier* (1957); *Tribe, Caste and Nation* (1960); *Politics and Social Change: Orissa in 1959* (1963); *Strategems and Spoils: A Social Anthropology of Politics* (1969); *Gifts and Poison: The Politics of Reputation* (editor, 1971); *The Politics of Innovation* (editor, 1973); *Morality and Expediency: The Folklore of Academic Politics* (1977); and *Tactical Uses of Passion: An Essay on Power, Reason and Reality* (1983).

A. P. BARNABAS, formerly, Professor of Sociology and Social Administration, Indian Institute of Public Administration, New Delhi. Author of *Social Change in a North Indian Village* (1967); *Citizens and Administration in a Developing Democracy* (co-author, 1969); *Population Control in India* (1978); and *Profile of the Child in India* (editor, 1981).

Y. B. DAMLE, Emeritus, Professor of Sociology, University of Poona. Author of *Communication of Modern Ideas and Knowledge in Indian Villages* (1955); and *Caste: A Trend Report and Bibliography* (co-author, 1959), *Intergroup Relations in Rural Communities of Maharashtra* (co-author, 1962); *College Youth in Poona: Elite in the Making* (1966); *Caste, Religion and Politics in India* (1982).

VICTOR S. D'SOUZA, formerly, Professor of Sociology, Punjab University, Chandigarh. Author of *Social Structure of a Planned City—Chandigarh* (1968); *Inequality and its Perpetuation: A Theory of Social Stratification* (1981); *Economic Development, Social Structure and Population Growth* (1985); and *Development Planning and Structural Inequalities: The Response of the Underprivileged* (1990).

RAJESH GILL, Senior Lecturer, Department of Sociology, Punjab University, Chandigarh. Author of *Social Change in Urban Periphery* (1991), *Slums as Urban Villages: A Comparative Study in Two Cities* (1994).

KHADIJA A. GUPTA, Reader in Sociology, Miranda House, University of Delhi. Author of *Politics in a Small Town* (1976); *Power Elite in India* (editor, 1989); and *Small Town Traders: A Sociological Study* (forthcoming).

WILLIAM H. NEWELL, formerly, Professor of Anthropology, University of Sydney. Author of *A Study of Gaddi—Scheduled Tribe—and Affiliated Castes* (1967); *Sociology of Japanese Religion* (editor, 1968); and *Ancestors* (editor, 1976).

M. N. PANINI, Professor of Sociology, Jawaharlal Nehru University, New Delhi. Co-author of *Basic Needs Viewed from Above and from Below—The Case of Karnataka* (1983); and *From the Female Eye: Accounts of Women Field Workers Studying their Own Communities* (editor, 1991).

E. A. RAMASWAMY, Professor of Industrial Relations, Institute of Social Studies, The Hague, Netherlands. Author of *The Worker and His Union : A Study in South India* (1977); *Industrial Relations in India* (1978); *Industry and Labour: An Introduction* (co-author, 1981); *Power and Justice: The State in Industrial Relations* (1984); *Worker Consciousness and Trade Union Response* (1988); and *Rayon Spinners: Strategic Management of Industrial Relations* (1994).

N. R. SHETH, formerly, Professor and Director, Indian Institute of Management, Ahmedabad. Author of *The Social Framework of an Indian Factory* (1968); *The Joint Management Council: Problems and Prospects* (1972); *Industrial Sociology in India* (co-author, 1979); and *Industrial Sociology in India: A Book of Readings* (editor, 1982).

HARSHAD R. TRIVEDI, Director, Institute of Cultural and Urban Anthropology, Ahmedabad. Author of *The Mers of Saurashtra*

(1961); *Urbanization and Macro-Social Change* (1975); *Urbanism: A New Outlook* (1976); *Housing and Community in Old Delhi* (1980); *Mers of Saurashtra Revisited* (1985); and *Mass Media and New Horizons: Impact of TV on Urban Milieu* (1991).

Index